British Poultry Standards

British Poultry Standards

Complete specifications and judging points of all standardized breeds and varieties of poultry as compiled by the specialist Breed Clubs and recognized by the Poultry Club of Great Britain

Fifth Edition

Edited by
Victoria Roberts
Council Member
Poultry Club of Great Britain

Blackwell
Science

Blackwell Publishing Ltd,
Editorial Offices:
Blackwell Science Ltd, 9600 Garsington Road, Oxford OX4 2DQ, UK
Tel: +44 (0)1865 776868
Iowa State Press, a Blackwell Publishing Company, 2121 State Avenue, Ames, Iowa 50014-8300, USA
Tel: +1 515 292 0140
Blackwell Science Asia Pty, 550 Swanston Street, Carlton, Victoria 3053, Australia
Tel: +61 (0)3 8359 1011

First published in 1954 by Iliffe Books, an imprint of the Butterworth Group
Second edition 1960
Third edition 1971
Reprinted 1976, 1977, 1978, 1980
Fourth edition 1982
Reprinted 1988
Reprinted by Blackwell Science 1994, 1995
Fifth edition 1997
Reprinted 1998, 2002, 2003

Library of Congress Cataloging-in-Publication Data

British poultry standards: complete specifications
and judging points of all standardized breeds and
varieties of poultry as compiled by the specialist
breed clubs and recognized by the Poultry Club
of Great Britain/edited by Victoria Roberts.—
5th ed.
p. cm.
ISBN 0-632-04052-1 (alk. paper)
1. Poultry—Judging. 2. Poultry breeds.
3. Poultry—Standards. I. Roberts, Victoria.
SF485.B75 1997
636.5′0021841—dc20 96-9548
CIP

ISBN 0-632-04052-1

A catalogue record for this title is available from the British Library

Set in 9.5/11pt Plantin
by DP Photosetting, Aylesbury, Bucks
Printed and bound in Italy using acid-free paper
by G. Canale & C., Turin, Italy

For further information on Blackwell Publishing, visit our website:
www.blackwellpublishing.com

Contents

Acknowledgements

The Poultry Club of Great Britain wishes to acknowledge the following for supplying illustrations. Nearly all of the 216 colour photographs were taken by John Tarren at the major shows, and we are grateful to him for his patience and perseverance. The Gold Partridge Dutch female was taken by Michael Corrigan, and those of the turkeys by Victoria Roberts. *Poultry World* supplied the photograph of the three light brown bantam eggs, the black and white photographs on pages 83, 135, 196, 197 and 239 are from A. Rice Poultry Photographs, and the illustrations on pages 94 and 95 are from the American Standard of Perfection, 1974.

The help of the Poultry Club Council and all the Breed Clubs is acknowledged, and much appreciated, in the compilation of this new edition.

Introduction

Who is this book for? Some may think it is only to inform poultry judges of the finer points of each breed, and of course this is a valuable function, but there is a wealth of information for the student of history, the conservationist, the exhibiting fancier and those who just want to keep hens or waterfowl as a hobby.

Breeders of commercial poultry rarely acknowledge their debt to the pure breeds, but they are only too willing to use a particular aspect, such as resistance to a certain disease, to increase the profitability of their commercial birds. This is only possible due to the dedication of fanciers in keeping bloodlines pure over many generations. Since all poultry are man-made, their ancestor being the Red Jungle Fowl (a galliform), it is possible, of course, to influence their type, characteristics and colour easily, which is why the commercial world made such enormous strides so quickly in being able to produce cheap and plentiful high protein food on demand from the 1950s onward. Such was the quantity of research done that the feeding of poultry then became a science and commercially produced, scientifically formulated rations made life much easier. Before the innovation of the hybrid – usually graced with a number rather than a name – commercial flocks consisted of Rhode Island Reds, white Leghorns, white Wyandottes and light Sussex for eggs, plus Indian Game crossed with white Sussex for meat. There are still some small flocks of these pure breeds being run commercially, ensuring that utility aspects are maintained, because in today's cost-conscious society there are not many who can afford to keep the purely decorative birds. Not all useful attributes consist of egg or meat production, however. The larger, maternal breeds such as the Cochins and Brahmas are valuable as broodies, geese as watchdogs, ducks as slug eaters. The foragers such as Leghorns and other light breeds keep insect numbers under control, remove weeds and, of course, provide rich nitrogenous fertilizer.

The popularity of poultry continues to increase, and even the newest and smallest farm parks and tourist attractions have a few fowl for added interest. When these are pure breeds, suitably labelled, it fuels the enthusiasm for others to take up the hobby. Not only is feeding made easier, there are many firms supplying suitable housing and equipment designed for the best welfare of the birds, and advances in veterinary research ensure that healthy stock is normal. Legislation concerning poultry tends to change with epidemics or scares but, in recent years, the Ministry of Agriculture has realized that legislation for commercial poultry has little relevance for the backyard breeder.

Records of poultry keeping go back centuries, but it is only since Victorian times that Standards have been written down for specific breeds. Survival of the fittest was definitely the main criterion in the past and breeds like the Old English Game fowl would have been bred true to type for hundreds of years. Five-toed fowls were mentioned in AD 50 and those with crests appeared in paintings and writings from the fourteenth century onwards. After cock-fighting was outlawed in England in 1849 the idea of exhibitions took root as a way of continuing the competition, but in a modified form.

The first Standards were produced in 1865 for just a handful of breeds to try to maintain uniformity; it was not until the turn of the century and the importation of breeds from the continent and America that a volume of any size appeared. The Poultry Club has always been the guardian of the Standards, but the Standards themselves are delineated

Plate 2

1 Standard hackle from barred Plymouth Rock male and similar breeds. Note the points of excellence – barring practically straight across feather, sound contrast in black and blue-white, barring and ground colour in equal widths, and barring carried down underfluff to skin. Tip of feather must be black.

1A Faulty saddle or neck hackle from similar variety. There is lack of contrast in barring, with dull grey ground colour and V-shaped bars.

2 Hackle as standard description from silver Campine, in which males are inclined to hen feathering. Note that the black bar is three times the width of ground colour and tip of feather is silver.

2A In this faulty hackle (also from silver Campine male) ground colour is too wide and barring narrow. Feather is without silver tip.

3 Standard hackle from Marans male. In this and some similar breeds evenness of banding is not essential, but it is expected to show reasonable contrast. It should, however, carry through to underfluff.

3A From the same group of breeds this feather is far too open in banding and lacks uniformity of marking. It is also light in undercolour.

4 Standard markings of female body feather in Plymouth Rocks and similar barred breeds where barring and ground colour are required to be of equal width. Note that barring runs from end to end of feather and that tip is black.

4A Faulty feather from same group. Note absence of barring to underfluff and V-shaped markings; also blurred and indistinct ground colour.

5 Sound body feather from silver Campine female showing standard silver tip and barring three times as wide as ground colour, as in the male. Gold Campine feathers are similar but for difference in ground colour.

5A Faulty female feather, again from silver Campine. Here again, as in 2A, barring is too narrow in relation to silver ground colour and tip of feather is black.

6 Body feather from Marans female, conforming to standard requirements. Note that the markings are less definite than in Rocks and Campines, and the black is lacking in sheen, while ground colour is smoky white.

6A Faulty Marans female feather. Lacks definition and contrast in banding, which is indefinite in shape, the blotchy ground colour making an indistinct pattern.

7 Excellent body feather from partridge Wyandotte female, showing correct ground colour and fine concentric markings. Note complete absence of fringing, shaftiness and similar faults. Fineness of pencilling is a standard requirement.

7A From the same breed this faulty female feather shows rusty red ground colour and indistinct pencilling, with faulty underfluff.

8 Body feather of standard quality from Indian or Cornish Game female. The illustration shows clearly two distinct lacings with a third inner marking. Lacing should have green sheen on a rich bay or mahogany ground.

8A Faulty feather from same breed. Missing are evenness of lacing and central marking. The outer lacing runs off into a spangle tip.

9 Standard feather from laced Barnevelder female. In this breed ground colour should be rich with two even and distinct concentric lacings. Quill of feather should be mahogany colour to skin.

9A Faulty Barnevelder female feather, showing spangle tip to outer lacing and irregular inner markings on ground colour that is too pale.

Plate 2

Plate 3

1 Standard markings on silver laced Wyandotte female feather, showing very even lacing on clear silver ground colour and rich colour in underfluff. In this breed clarity of lacing is of greater importance than fineness of width.

1A Faulty female feather from same breed. In this there is a fringing of silver outside the black lacing, which is irregular in width and runs narrow at sides. Undercolour is also defective.

2 Excellent feather from gold laced Wyandotte. In this ground colour is a clear rich golden-bay and there is a complete absence of pale shaft. Undercolour is sound and lacing just about the widest advisable.

2A This shows a very faulty feather from same breed. It portrays mossy ground colour with blotchy markings and uneven width of lacing at sides of feather. Undercolour is not rich enough.

3 Standard markings on Andalusian female feather showing well-defined lacing on clear slate-blue ground and good depth of colour in underfluff. The dark shaft is desirable and is not classed as a fault.

3A Faulty feather from female of same breed. In this the ground colour is blurred and indistinct, and the lacing is not crisp, while undercolour lacks depth.

4 This shows a feather from an Ancona female, almost perfect in standard requirements. The white tipping is clear and V-shaped and undercolour is dark to skin.

4A Faulty feather from female of same breed. Here the tip of feather is greyish-white and lacks the necessary V-shape, while undercolour is not rich enough.

5 An almost perfectly marked feather from a speckled Sussex female – though the white tip might be criticized by some breeders as rather too large. The black dividing bar shows good green sheen and the ground colour is rich and even.

5A As a contrast this faulty feather shows a blotchy white tip and lack of colour in underfluff. The ground colour is also uneven.

6 An excellent example of 'mooning' on the feather of a silver spangled Hamburgh female. Note the round spangle and the clear silver ground with sound undercover.

6A In this feather from the same breed the spangling at tip is not moon-shaped and there is too much underfluff and insufficient silver ground colour to body of feather.

7 A good example of the desired colour in Rhode Island Red female plumage. Note the great depth of rich colour and the sound dark undercolour.

7A Faulty colour in a feather from the same breed. Here the middle of feather is paler and inclined to shaftiness, and colour generally is uneven.

8 Standard plumage in females of Australorp and similar breeds of soft feather with rich green sheen. Note the brilliance of colour and general soundness of underfluff.

8A This shows a common fault in similar breeds, a sooty or dead black colour without sheen and lacking lustre. This sootiness is, however, usually accompanied by dark undercolour.

9 Standard colour and feather in the buff Rock female and similar breeds which perhaps vary in exact shade and in quantity and softness of underfluff. Note the clear even buff and lack of shaftiness or lacing, also the sound rich undercolour.

9A This feather from a similar buff breed shows very bad faults – mealiness and bad undercolour with a certain amount of pale colour in shaft.

Plate 3

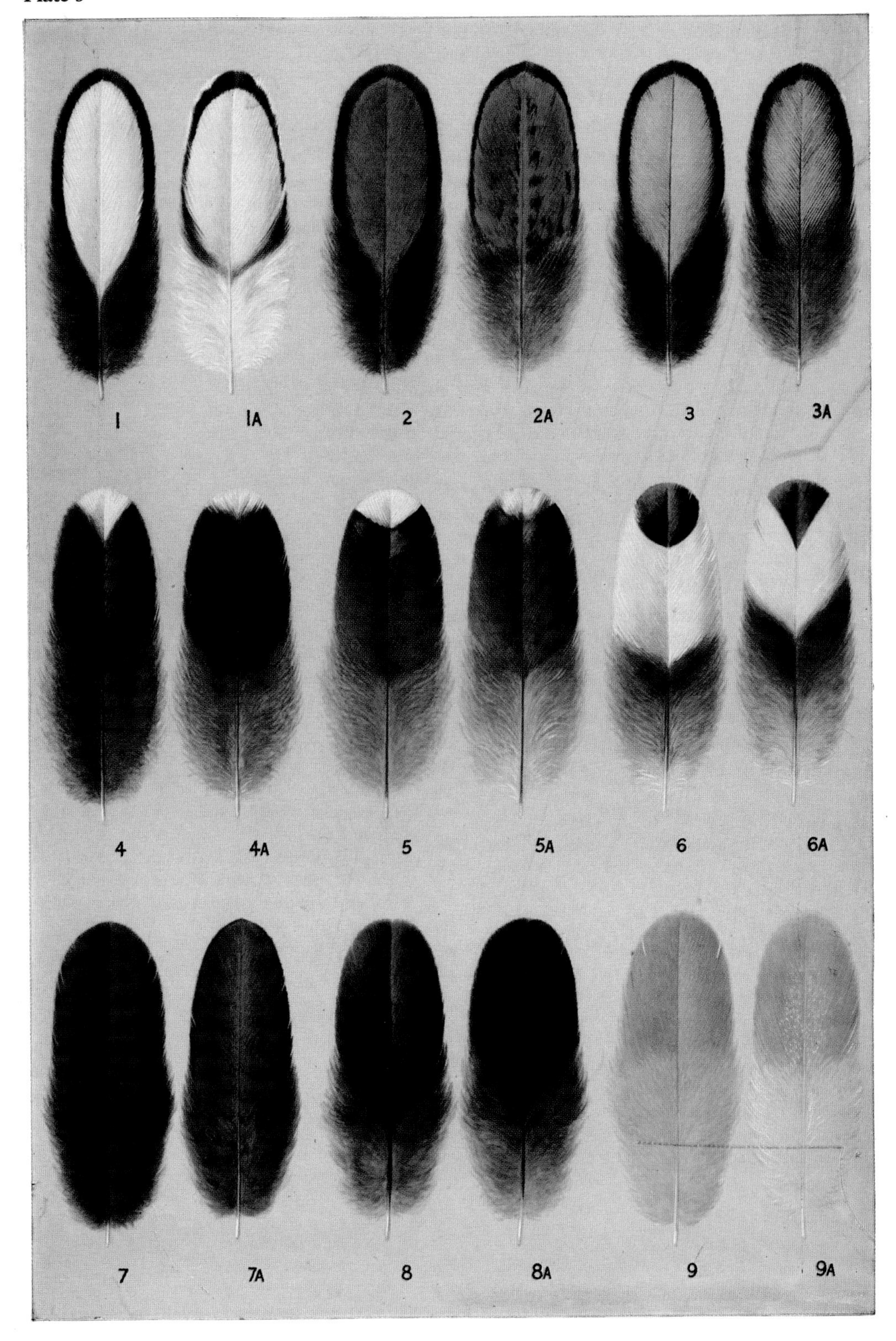

Plate 4

1 This shows a typical standard bred feather from a Derbyshire Redcap female. Note the rich ground colour and the crescentic black markings, which are really midway between spangling and lacing.

1A In this faulty feather from a female of the same breed the ground colour is uneven and lacks richness, while the black tip is too small and indefinite and too closely resembles moon-shaped spangling.

2 This is a standard example of the webless type of plumage associated with Silkies in which the feather vane has no strength and the barbs no cohesion. This plumage is common in all colours.

2A Faulty feather from the same breed. In this the middle of feather is too solid and lacks silkiness, while the fluff has insufficient length.

3 A delicately pencilled body feather from a silver grey Dorking female. Note the silvery colour and absence of ruddy or yellow tinge in ground colour. This type of feather is also usual in duckwing females of various breeds.

3A Faulty colour in female feather from same breed. Here there is a distinctly incorrect ground colour and pronounced shaftiness.

4 A good example of standard bred colour and markings in body feather of brown Leghorn female, where the ground colour is a soft brown shade and the markings finely pencilled. This type of feather is common to many varieties of partridge or grouse colouring.

4A This shows a body feather from the same breed, in which ground colour is ruddy and shaftiness is pronounced – both severe exhibition faults.

5 A well-chosen example of the irregularity in markings of an exchequer Leghorn female. In this breed the black and white should be well distributed but not regularly placed, and underfluff should be parti-coloured black and white.

5A This faulty feather from the same breed shows a too regular disposition of markings, the body of the feather being almost entirely black and the white markings almost resembling lacing.

6 This is a standard feather from the breast of a silver Dorking, and with slight variations of shade from pale to rich salmon applies to a number of varieties with black-red or duckwing colouring. Colour should be even with as little pale shaft as possible.

6A A faulty sample of breast feather from the same group. Here the ground colour is washy and disfigured by pale markings known as mealiness.

7 Standard markings in North Holland Blue female. Note the defined but somewhat irregular banding on a distinctly bluish ground. No banding or other requirements in underfluff are called for in the standard.

7A This shows a faulty female feather in the same breed, which is not closely standardized for markings. The ground colour is smoke-grey instead of blue, and is blotchy, with uneven markings.

8 A good example of clear colour in an unlaced or self-blue female feather, where no lacing is permissible, such as in blue Leghorns, blue Wyandottes, etc. Note even pale blue shade and absence of any form of markings. This is an example of the true-breeding blue colour found in Belgian bantams.

8A This faulty female feather is a dull dirty grey instead of clear blue, and has blotchy markings as well as a suggestion of irregular lacing.

9 A good sample of exquisitely patterned thigh fluff in Rouen drakes. The ground colour is a clear silver and the markings a delicate but clear black or dark brown. These markings are sometimes known as chain mail.

9A Another good Rouen feather – this time from the duck. Ground colour is very rich and markings intensely black, though seldom so regular and even as in domestic fowl.

Plate 4

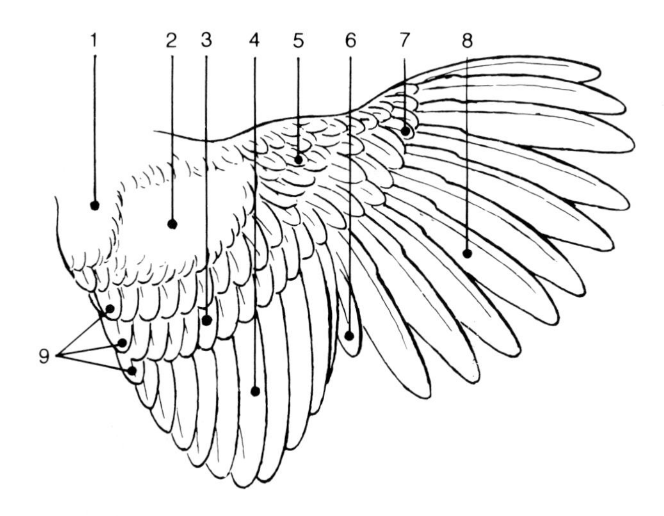

Figure 2

1 Shoulder butt or scapulars
2 Wing bow coverts
3 Wing bar or speculum (lower wing coverts)
4 Secondaries
5 Wing bow coverts
6 Axial feather (not waterfowl)
7 Flight coverts
8 Primaries
9 Tertiaries (mainly waterfowl)

Figure 3 Types of comb

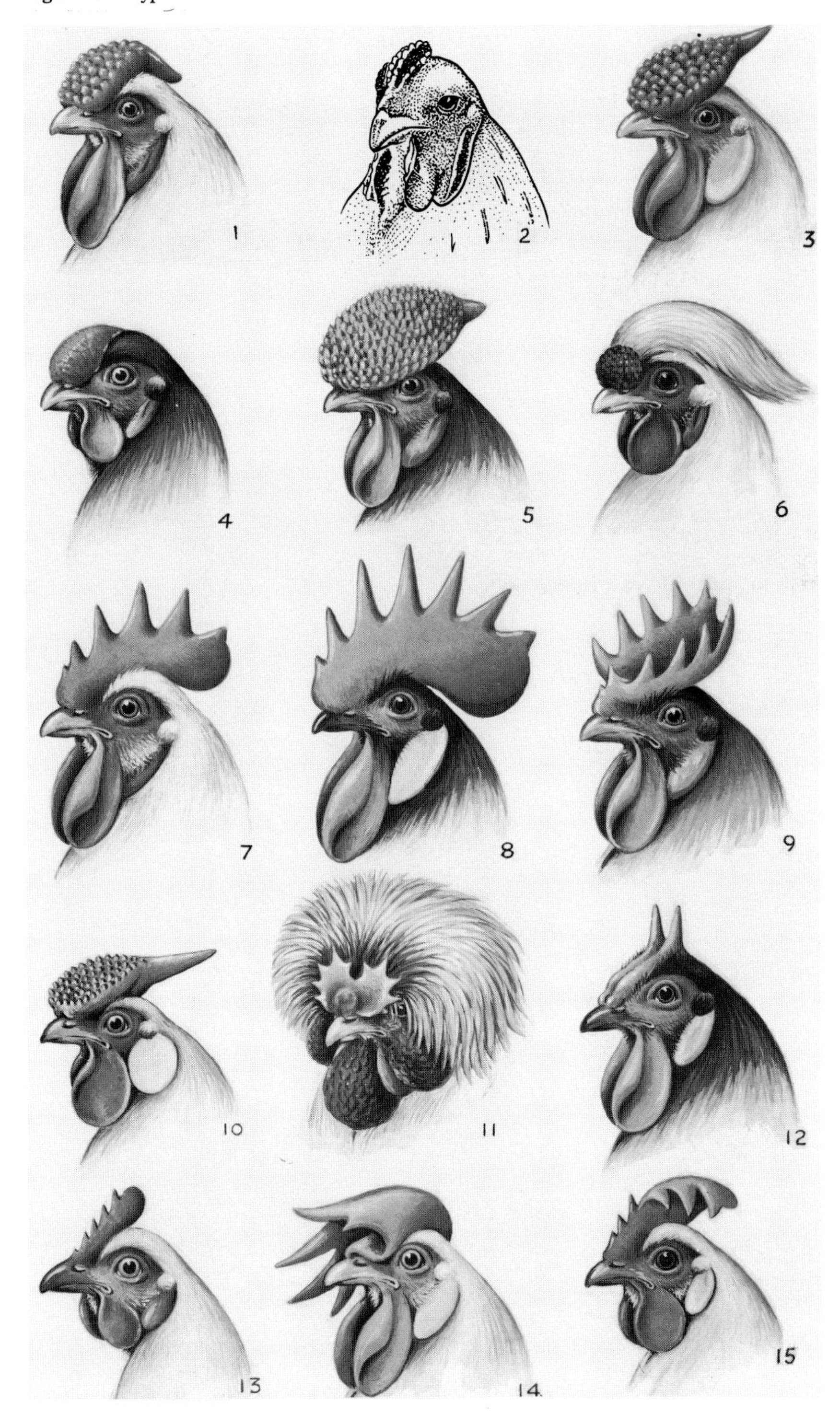

1 Rose, leader following line of neck **2** Triple or pea **3** Rose, short leader **4** Walnut **5** Cap **6** Mulberry **7** Medium single **8** Large single **9** Cup **10** Rose with long leader **11** Leaf **12** Horn **13** Small single **14** Folded single **15** Semi-erect single

Figure 4 Leg types

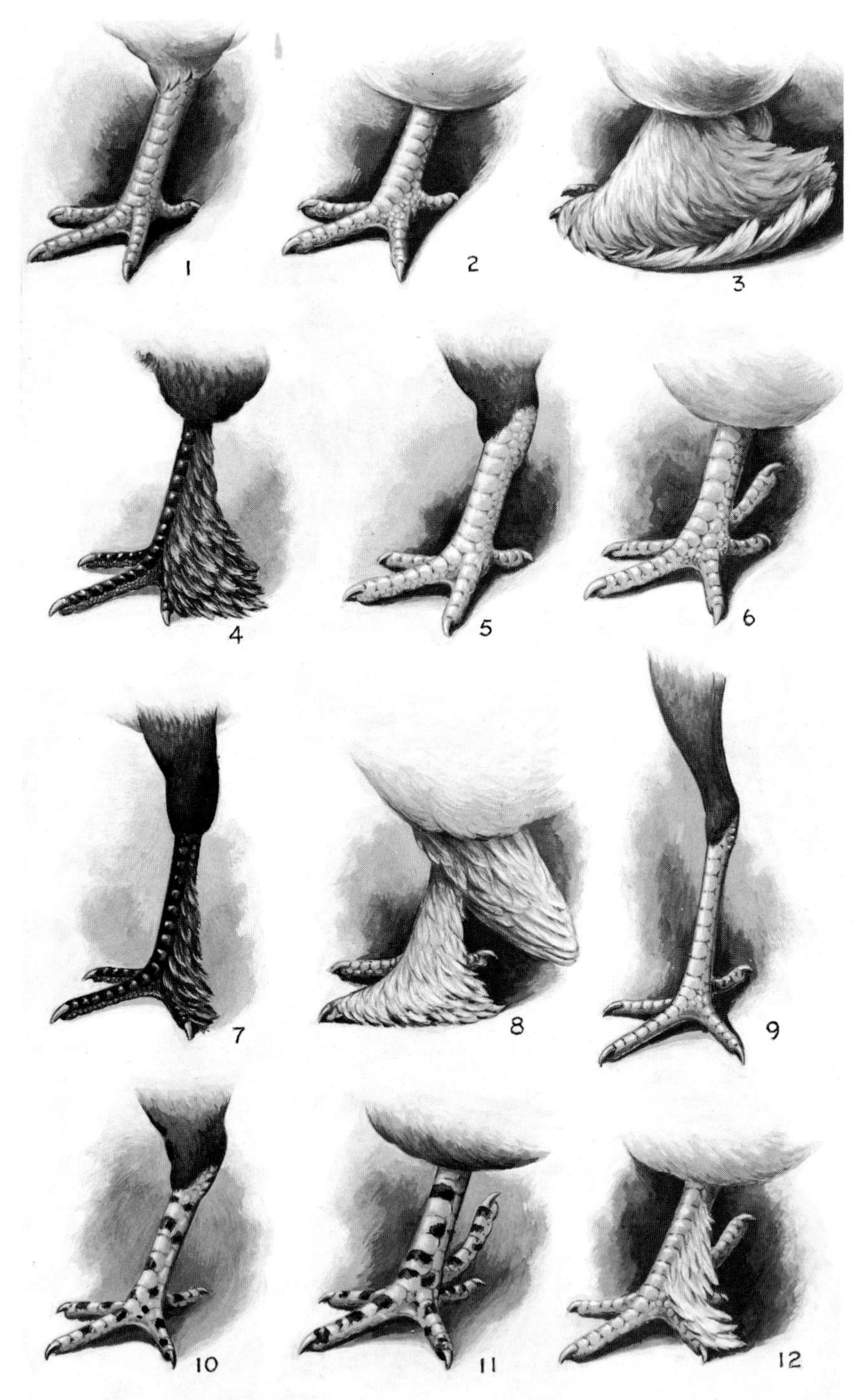

1 Clean legged, flat side (Leghorns) **2** Clean legged round shanks (Game) **3** Heavy feather legged, and feathered toes, i.e. foot feather **4** Feather legged, no feathers middle toe (Croad Langshan) **5** Short round shanks (Indian Game) **6** Five toed (Dorking) **7** Slightly feathered shanks (Modern Langshan) **8** Feather legged and vulture hocked **9** Thin round shanks (Modern Game) **10** Mottled shanks (Ancona) **11** Mottled and five toed (Houdan) **12** Feather legged and five toed (Faverolles)

Figure 5 Types of tail

Figure 6

1 Neck hackle, male (striped)
2 Neck hackle, female (laced)
3 Saddle hackle, male (striped)
4 Pencilled hackle (female)
5 Ticked hackle
6 Tipped neck hackle, male, as in spangled Hamburgh
7 Striped hackle, male. Shows outer fringing of colour – a fault
8 Striped saddle hackle, male, showing open centre (desired only in pullet-breeder)
9 Pencilled feather, cushion, female, as in silver grey Dorking and brown Leghorns
10 Barred neck hackle (male)
11 Triple pencilled back (female)
12 Laced
13 Faulty laced (i.e. horseshoed)
14 Spangled (moon-shaped)
15 Speckled. Irregular-shaped white tick shows three colours on feather
16 Shoulder feather in spangled varieties
17 Poland laced crest (pullet)
18 Poland crest, female
19 Crescent marked
20 Barred or finely pencilled as in Hamburgh. Bars and spaces same width
21 Double laced
22 Tipped, showing 'V'-shaped tip, as in Ancona
23 Barred as in barred Rock, shows barring in undercolour. To finish with black bar
24 Laced and ticked, as in dark Dorking
25 Elongated spangle, as in Buttercup
26 Finely pencilled, as in dark Brahma female
27 Barred, as in Campine. Finishes with white end. Light bars a quarter to a third of the width of dark bars
28 'Silkie' (no webbing)
29 Fine in pencilling, as in black marks of black-red, and duckwing Game
30 Barred Rock sickle
31 Buff laced
32 Wing marking on flight feather
33 Laced sickle
34 Saddle hackle mackerel marked (Campine cockerel)

Figure 6 Feather markings

Classification of breeds

POULTRY

Hard Feather

Asil (Rare)
Belgian Game (Rare)
Indian Game
Ko-Shamo (Rare)
Malay (Rare)
Modern Game
Nankin-Shamo (Rare)
Old English Game Bantam
Old English Game Carlisle
Old English Game Oxford
Rumpless Game (Rare)
Shamo (Rare)
Tuzo (Rare: True Bantam)
Yamato-Gunkei (Rare)

Soft Feather: Heavy

Australorp
Barnevelder
Brahma
Cochin
Croad Langshan
Dorking
Faverolles
Frizzle
Marans
Orpington
Plymouth Rock
Rhode Island Red
Sussex
Wyandotte

Soft Feather: Light

Ancona
Appenzeller
Araucana
Rumpless Araucana
Hamburgh
Leghorn
Minorca
Poland
Redcap
Scots Dumpy
Scots Grey
Silkie
Welsummer

True Bantam

Belgian
Booted (Rare)
Dutch
Japanese
Nankin (Rare)
Pekin
Rosecomb
Sebright
Tuzo (Hard Feather: Rare)

Rare

Hard Feather

Asil
Belgian Game
Ko-Shamo Bantam
Malay
Nankin-Shamo Bantam
Rumpless Game
Shamo
Tuzo (True Bantam)
Yamato-Gunkei

Soft Feather: Heavy

Autosexing breeds:
  Rhodebar
  Wybar
Crèvecoeur
Dominique
German Langshan
Houdan
Ixworth
Jersey Giant
La Flèche
Modern Langshan
New Hampshire Red
Norfolk Grey
North Holland Blue
Orloff
Transylvanian Naked Neck
Turkeys

Rare

Soft Feather: Light		True Bantam
Andalusian	Kraienköppe	Booted
Augsberger	Lakenvelder	Nankin
Autosexing breeds:	Marsh Daisy	Tuzo (Hard Feather)
Legbar	Old English Pheasant Fowl	
Welbar	Sicilian Buttercup	
Brakel	Spanish	
Breda	Sulmtaler	
Campine	Sultan	
Fayoumi	Sumatra	
Friesian	Vorwerk	
Italiener	Yokohama	

DUCKS

Heavy	Light	Bantam
Aylesbury	Abacot Ranger	Black East Indian
Blue Swedish	Bali	Call
Cayuga	Campbell	Crested
Muscovy	Crested	Silver Appleyard Miniature
Pekin	Hook Bill	Silver Bantam
Rouen	Indian Runner	
Rouen Clair	Magpie	
Saxony	Buff Orpington	
Silver Appleyard	Welsh Harlequin	

GEESE

Heavy	Medium	Light
African	American Buff	Chinese
Embden	Brecon Buff	Pilgrim
Toulouse	Buff Back	Roman
	Grey Back	Sebastopol
	Pomeranian	Steinbacher

SITTERS AND NON-SITTERS

Generally speaking these divide themselves – *heavy* breeds being sitters and the *light* breeds non-sitters – the former comprising mainly American and Asiatic breeds with Indian Game, Sussex, and Dorkings of British origin, while the latter are generally of Mediterranean origin. Unlike certain other countries, however, which classify three categories – *light*, *medium* and *heavy* – Great Britain adheres to two classes only. There are, therefore, certain exceptions to the foregoing generalization. The breeds are divided as follows:

Sitters

Asil
Australorp
Barnevelder
Brahma
Cochin
Crèvecœur
Croad Langshan
Dorking
Faverolles
Frizzle
Houdan
Indian Game
Ixworth
Jersey Giant
Jubilee Indian Game
La Flèche
Malay
Marans
Marsh Daisy
Modern Game
Modern Langsham
New Hampshire Red
Norfolk Grey
North Holland Blue
Old English Game
Orloff
Orpington
Plymouth Rock
Rhode Island Red
Scots Dumpy
Silkie
Sultan
Sumatra Game
Sussex
Wyandotte
Yokohama

Non-sitters

Ancona
Andalusian
Bresse
Campine
Hamburgh
Lakenvelder
Leghorn
Minorca
Old English Pheasant Fowl
Poland
Redcap
Scots Grey
Sicilian Buttercup
Spanish
Welsummer

Bantams take the same classification, i.e. heavy or light, as their large prototypes.

Defects and deformities

The Poultry Club instructs its judges to work continuously for stock improvement in making their awards. They should keep in mind, therefore, the suitability of exhibits for the breeding pen, penalizing those defects that affect reproductive values or detract from what may be regarded as the highest merits of such birds.

Uniformity in judging and exhibiting is sought, the aim being to make show-pens reliable 'shop windows', displaying birds that buyers may claim with every confidence in their soundness and reproductive ability.

TO BE PASSED OR PENALIZED

The following are given as deformities and defects for which judges must pass or penalize an exhibit according to the seriousness of the defect.

Head points

Crossed or deformed beak. Malformation of beak. Badly dished bill in ducks. Blindness. Defective eyesight. Defective pupils. Odd eye colour. Comb that closes the nostrils. Side sprigs or double end on a single comb. Excessive fall of rose comb on a single base. Split combs at blade. Fall-over comb that obstructs vision. Malformed combs. Defective serrations. Peculiar head carriage. White in face. Wry neck. Indications of brain or nerve affection. Badly distended and sagging crop.

Crossed beak　　Open beak　　Sunken eye

Figure 7

Figure 8 Dished bill

Back

Any deformity. Rounded or curved spine (roach back). Weak back formation. One bone higher than the other giving the back a lopsided appearance.

Bone structure

Pigeon breast (abnormal protrusion of the breastbone). Seriously deformed breastbone. Malformation of breastbone that interferes with the internal organs. Down behind and curved end of breastbone which leads to drooping abdomen. Dented breastbone from perching. Enlargement on breastbone of turkey. Broken or malformed pelvic bones. Faulty stance.

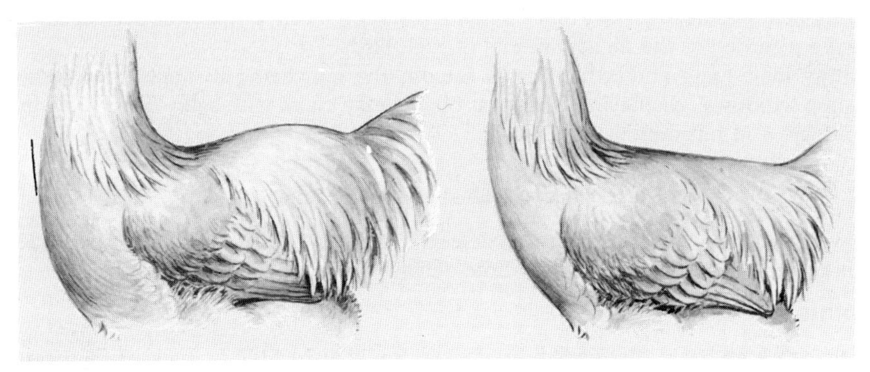

Roach back — Cut-away breast

Figure 9

Wings

Badly twisted or curled wing feathers. Slipped or drooping wing. Split wing, in a serious form, with large gap between primaries and secondaries. Defective wing formation in waterfowl. Slightly defective wing formation, even if well positioned and carried, to be penalized.

Tail

Wry. Squirrel. Defective parson's nose. Split or divided tail feathers or badly twisted feathers in tail. High tail in excess.

Legs and feet

Enlarged bone. Curved thigh bones. Malformation of bone. Bow legs or 'out of hocks'. Badly in at hocks. Duck toes. Crooked toes. Turned toes. Twisted feet. Enlarged toe joints. Lack of spurs on adult male. Leg feathers on clean legged birds.

Feathering

Soft or frizzled feathering in plain feathered breeds. Curled feathers on any part of body, including neck. Signs of slow feathering.

Disease

Any disease or disorder not making for maximum health, condition, vigour and breeding fitness, including colds, heartiness, abdominal dropsy, cysts, egg substance in oviduct or abdomen. Sour crop. Impacted crop. Any other disease symptom or deformity.

Figure 10 Comb faults

Left Ingrown leader

Right Short of leader, and uneven wattles

Left Rose comb falling to side and blocking vision

Right Bad leader and coarse worked comb

Left Beefy, and with part of blade too far forward

Right Badly curved at rear end with spikes falling over

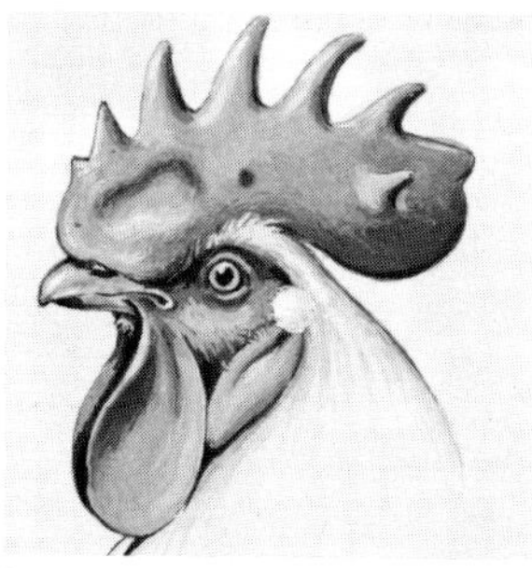

Left Thumb mark and side sprig

Right Flyaway comb

Left Double folded comb

Right Flop comb blocking vision

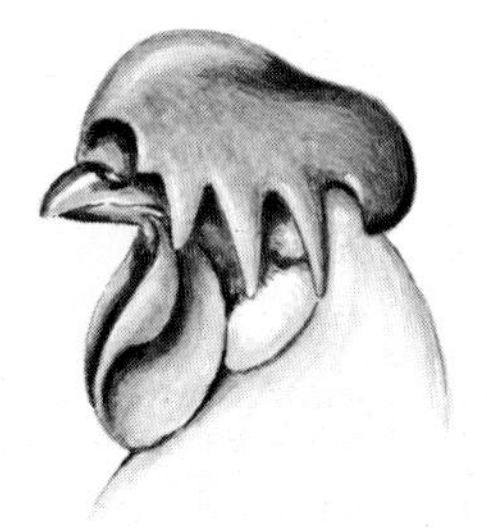

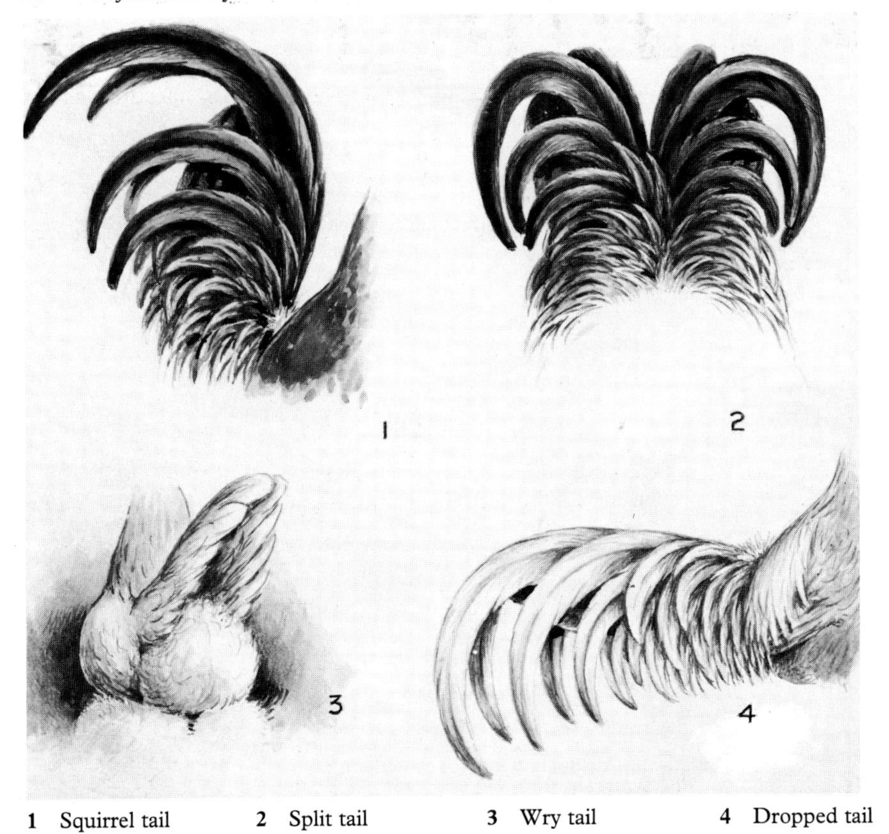

1 Squirrel tail 2 Split tail 3 Wry tail 4 Dropped tail

Figure 11

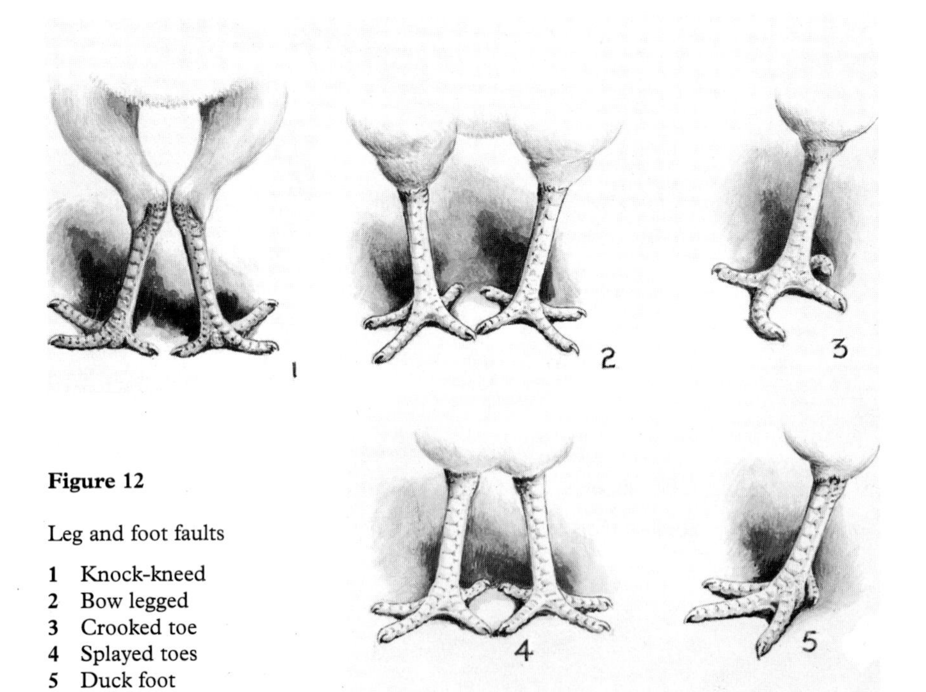

Figure 12

Leg and foot faults

1 Knock-kneed
2 Bow legged
3 Crooked toe
4 Splayed toes
5 Duck foot

Lack of breed characteristics

Any exhibit deficient in breed characteristics, so that it is an unworthy speciman of the breed or variety intended, must be passed.

DISQUALIFICATIONS

Any bird that in the opinion of the judge has been faked or tampered with shall be disqualified.

Large fowl and bantams

ANCONA

LARGE FOWL

Origin: Mediterranean
Classification: Light: Soft feather
Egg colour: White to cream

Named after the province of Ancona in Italy, specimens of this Mediterranean breed were imported into England in 1851, first the single then the rose comb. Controversy centres around the view that Anconas are akin to the original mottled Leghorn and, therefore, a member of the Leghorn family. However, the fact remains that breeders adhere to the name of Ancona. The breed has retained its popularity on the show-bench not only for its laying propensities, but because of its combination of breed type and characteristics with usefulness.

General characteristics: male

Carriage: Alert, bold and active.
Type: Body broad, close and compact. Back of moderate length. Breast full and broad, carried well forward and upward. Wings large and carried well tucked up. Tail full and carried well out.
Head: Deep, moderate in length, rather inclined to width, and carried well back. Beak medium with a moderate curve. Eyes bright and prominent. Comb single or rose. The single of medium size, upright and with five to seven deep, broad and even serrations forming a regular curve, coming well back and following the line of the head, free from excrescences. The rose resembles that of the Wyandotte. Face smooth and of fine texture. Ear-lobes medium, inclined to almond shape, free from folds. Wattles long, fine in texture, in proportion to comb.
Neck: Long, nicely arched and well covered with hackle.
Legs and feet: Legs of medium length, strong, set well apart, clear of feathers, thighs not much seen. Toes, four, rather long and thin and well spread out.

Female

With the exception of the single comb, which falls gracefully to one side of the face without obscuring the vision, the characteristics are generally similar to those of the male, allowing for the natural sexual differences. The body, however, is round and compact, with greater posterior development than in the male. The back is rather long and broad, the neck of medium length and carried well-up.

Colour

Male and female plumage: Good beetle-green ground tipped with white (the more V-shaped the better). No inclination to lacing. The more evenly V-tipped throughout with beetle-green and white the better, provided the ground colour is beetle-green.

In both sexes: Beak yellow with black or horn shadings; a wholly yellow beak not desirable. Eyes, iris orange-red, pupil hazel. Comb, face and wattles bright red, face free

Ancona male, large

Ancona female, bantam

from white. Ear-lobes white. Legs yellow, mottled with black, the more evenly mottled the better.

Weights

Cock	2.70–2.95 kg (6–6½ lb);	cockerel	2.50 kg (5½ lb)
Hen	2.25–2.50 kg (5–5½ lb);	pullet	2.00 kg (4½ lb)

Scale of points

Type and carriage	15
Texture (general)	10
Size	5
Purity of white, quality and evenness of tipping	20
Beetle-green ground colour dark to skin	15
Leg colour	5
Head (eye 5, comb 10, lobe 5)	20
Beak colour	5
Condition	5
	100

(*Note:* Eye points include brightness and prominence. Comb points include medium size and fine texture.)

Defects

	To lose
In-kneed	10
Squirrel tail	10
Crooked toes	10
White or light undercolour	10
Ground colour other than beetle-green	10
Tail not tipped or not black to roots	10
Wings any other colour than black tipped with white	10
Bad comb	5
White in face	20
Lobe other than white	5
	100

(*Note:* Roach back or any bad structural deformity a disqualification.)

BANTAM

Ancona bantams are to be exact miniatures of their large fowl counterparts and so standard, colour and scale of points are to be used as for large fowl.

Weights

Male	570–680 g (20–24 oz)
Female	510–620 g (18–22 oz)

ANDALUSIAN

LARGE FOWL

Origin: Mediterranean
Classification: Light: Rare
Egg colour: White

The breed owes its name to the Province of Andalusia in Spain, and is one of the oldest of the Mediterranean breeds. It is a contemporary of the black Spanish to which, no doubt, it is closely related. The blue Andalusian, as we know it today, was developed from black and white stock imported from Andalusia about 1846, and blending of the two colours most probably created the blue. The earlier specimens were large and game-like in carriage, with medium combs and lobes, and of a self colour, although individual birds were selectively bred for lacing by infusion of black Minorca blood.

General characteristics: male

Carriage: Upright, bold and active.
Type: Body long, broad at the shoulders, and tapering to the tail, with the plumage close and compact. Breast full and round. Wings long, well tucked up and the ends covered by the saddle hackles. Tail large and flowing, carried moderately high but not approaching 'squirrel' or fan shape.
Head: Moderately long, deep and inclined to width. Beak stout and of medium length. Eyes prominent. Comb single, upright and of medium size, deeply serrated with spikes broad at the base, the back portion slightly following the line of the head but not touching the neck. Free from 'thumb marks' or side spikes. Face smooth. Ear-lobes almond in shape, medium size, free from wrinkles, and fitting closely to the face. Wattles fine and long.
Neck: Long and well covered with hackle feathers.
Legs and feet: Legs long. Shanks and feet free from feathers. Toes, four, straight and well spread.

Female

With the exception of the comb, which falls with a single fold to one side without covering the eye, the general characteristics are similar to those of the male, allowing for the natural sexual differences.

Colour

Male and female plumage: Clear blue, edged with distinct black lacing, not too narrow, on each feather, excepting the male's sickles, which are dark (or even black), and his hackles, which are black with a rich gloss, while the female's neck hackle is a rich lustrous black, showing broad lacing on the tips of the feathers at the base of the neck. Undercolour to tone with surface colour.

In both sexes: Beak dark slate or horn. Eyes dark red or red-brown. Comb, face and wattles bright red. Ear-lobes white. Legs and feet dark slate or black.

Weights

Male	3.20–3.60 kg (7–8 lb)
Female	2.25–2.70 kg (5–6 lb)

Scale of points

Ground colour	30
Lacing	20
Head (comb 10, face 10, lobes 5)	25
Size, type, carriage, tail and condition	25
	100

Serious defects

In the male, much white in face or presence of red in lobes. White feathers. Sooty ground colour. Red or yellow in hackles. Any deformity and comb not upright. In the female any of these points that apply, together with an upright comb.

Andalusian male, bantam

Andalusian female, bantam

BANTAM

Andalusian bantams are exact miniatures of their fowl counterparts and so standard, colour and scale of points apply.

Weights

Male 680–790 g (24–28 oz)
Female 570–680 g (20–24 oz)

APPENZELLER

LARGE FOWL

Origin: Switzerland
Classification: Light: Soft feather
Egg colour: White

The two types of Appenzeller have recently been imported into this country from the Continent. They have been popular and have attracted enough supporters to warrant forming a specialist club to look after the interests of both the Spitzhauben and the Barthuhner. There is no Appenzeller bantam standardized at the moment.

The Spitzhauben

General characteristics: male

Carriage: Neat, and very active.
Type: Body well rounded, medium long, walnut-shaped. Breast full, carried high. Wings rather long, carried close. Tail well furnished, well spread, at right angles to back. Abdomen well developed. Back of medium length, slightly sloping with full hackle.
Head: Medium sized, held high, with typical raised skull and medium-sized pointed crest bent forward. Face smooth. Comb horn type consisting of two small rounded spikes, separate and without side sprigs. Wattles moderately long and fine. Ear-lobes medium sized, oval. Beak powerful, with strong, cavernous nostrils, and a prominent horseshoe ridge to the beak with a small fleshy knob at the front. Eyes prominent and alert.
Neck: Medium length, slightly arched with abundant hackle.
Legs and feet: Thighs slender and prominent; shanks medium length, fine. Toes, four, well spread.
Plumage: Fairly hard and tight.

Female

Except for a more horizontal back line the general characteristics are the same for the male, allowing for the natural sexual differences.

Colour

The silver spangled

Male plumage: Pure silvery white ground colour, with each feather ending in distinct black, fairly small spangle, not circular; less pronounced on the head and neck. Primaries, secondaries and tail feathers with black tips. Abdomen and fluff grey; underplumage dark grey.
Female plumage: Head crest and the neck silver-white with black tipping. Breast, wing bows, back and tail silvery white with distinct black spangling. Flights as for the male. Undercolour dark grey.

The gold spangled

Male plumage: Gold-red ground colour, spangling as for the silver spangled. Flights: outer web golden-yellow, inner web as black as possible. Breast and flanks gold with black spangles. Abdomen and underplumage greyish black. Tail as brown as possible with black tips, a blackish brown tail allowed.
Female plumage: Golden-yellow ground colour, tail golden-brown with black spangling. Otherwise as for the male, having regard to the necessary sexual differences.

The black

Male and female plumage: Shiny greenish-black with dark grey to black underplumage.

In both sexes and all colours

Beak bluish. Eyes dark brown. Comb, face and wattles bright red; ear-lobes bluish-white. Shanks blue.

Weights

Male	1.60–2.00 kg ($3\frac{1}{2}$–$4\frac{1}{2}$ lb)
Female	1.35–1.60 kg (3–$3\frac{1}{2}$ lb)

Scale of points

Type and carriage	25
Colour and markings	25
Head points	25
Legs and feet	15
Condition	10
	100

Serious defects

Comb other than horn. Side sprigs. Narrow or roach back. Squirrel tail. Breast too deep or narrow. Low wing carriage. Tail lacking fullness. Crow beak. Nostril not cavernous. Bad stance. Any sign of feathering on shanks.

The Barthuhner

General characteristics: male

Carriage: Strong and active.
Type: Body well rounded, medium long with broad shoulders. Breast full, carried high. Wings moderate, tucked in close to the body. Tail well furnished, carried like a fan with abundant sickle feathers. Back of medium length, slightly sloping with full saddle hackle.
Head: Medium sized with medium-sized rose comb with good, rounded working and a straight leader. Wattles small, covered by a full cheek and chin beard. Large beak and nostrils. Eyes prominent and alert.
Neck: Medium length, slightly arched with abundant hackle.
Legs and feet: Thighs well developed, shanks of medium length with no feathering. Four toes, well spread.
Plumage: Fairly tight.

Female

Except for a more horizontal back line, the general characteristics are the same as for the male, allowing for the natural sexual differences.

Colour

As in any of the recognized game fowl. The blue colour as per the Orpington standard.

Silver Appenzeller Spitzhauben male

Silver Appenzeller Spitzhauben female

In both sexes and all colours

Comb and wattles red, ear lobes white, beak black or horn, eyes as dark as possible, legs blue to black.

Eggs

Tinted

Weights

Male 2.26–2.94 kg (5–6½ lb)
Female 1.36–1.81 kg (3–4 lb)

Scale of points

Type and carriage	30
Colour	20
Head points	30
Legs and feet	10
Condition	10
	100

Serious defects

Comb other than rose. Narrow or roach back. Squirrel tail. Missing beard or any sign of feathering on shanks.

ARAUCANA

LARGE FOWL

Origin: Chile
Classification: Light: Soft feather
Egg colour: Blue or green

When the Spaniards arrived in South America, bringing with them the light Mediterranean breeds, they found that the indigenous Indians had domestic fowl which soon cross-bred with the incomers. Notable for their fierce resistance to the Spaniards, however, were the Indians of the Arauca province of northern Chile who were never conquered. The name Araucana for the breed is derived therefore from that part of the world where the South American and European fowls had the least opportunity to interbreed.

The Araucana breed standard in the British Isles is generally as envisaged by George Malcolm who created the true-breeding lavender Araucana, among other colours, in Scotland during the 1930s. Araucanas are prolific layers of strong-shelled eggs, blue or green eggs having been reported from South America from the mid-sixteenth century onwards. These are unique in that their colour permeates throughout the shell.

General characteristics: male

Carriage: Alert and active
Type: Body long and deep, free from heaviness. Firm in handling. Back moderately long, horizontal. Wings large and strong. Tail well developed with full sickles carried at an angle of 45°.
Head: Moderately small. Beak strong and stout. Eyes bold. Comb small pea. Face covered with thick muffling and ear muffs abundant. Crest compact, carried well back from eyes. Ear-lobes moderately small and concealed by muffling. Wattles absent.
Neck: Of medium length abundantly furnished with hackle feathers.

Pyle Araucana male, large

White Araucana female, bantam

Lavender Araucana female, bantam

Legs and feet: Medium length, strong and well apart. Shanks free from feathers. Toes, four straight and well spread.

Female

The general characteristics are similar to those of the male allowing for the natural sexual differences. Comb pea.

Colour

The lavender

Male and female plumage: An even shade of blue-grey throughout.

The blue

Male plumage: Breast, belly, thighs, tail and closed secondaries the colour of new slate. Hackle, saddle and shoulders, and sometimes the tail coverts and the primaries, two shades darker (like a slate after being wetted). Fluff slate-blue.

Female plumage: Blue slate colour with dark hackle like the male, often marked or laced all over with the darker shade. Fluff slate-blue.

The black-red

Male plumage: Breast, thighs, belly, tail and wings black. Wing bars green-black; secondaries when closed bay. Crest, head and neck orange-red striped black. Back, shoulders and wing bow red or mahogany. Saddle hackle to match neck hackle. Fluff grey.

Female plumage: Hackle rich golden yellow broadly striped with black. Breast salmon. Muff salmon. Salmon and ash-grey on thighs. Body colour brown pencilled black, each feather with a pale shaft. Tail brown spotted or grizzled with black. Fluff grey.

The silver duckwing

Male plumage: Resembles the black-breasted red in the black markings and blue wing bars; rest of the plumage clear silvery white. Fluff light grey.

Female plumage: Hackle white, lightly striped black. Body and wings even silvery grey. Breast pale salmon. Primaries and tail nearly black. Fluff light grey.

The golden duckwing

Male plumage: Hackle and saddle yellow-straw. Shoulders deep golden. Wing bars steel blue; secondaries yellow or creamy straw when closed, remaining plumage black. Fluff light grey.
Female plumage: Breast deeper, richer colour and body slightly browner tinge than the silver duckwing female. Fluff light grey.

The blue-red

Male plumage: The same colour pattern as the black-red with slate replacing black. Breast, thighs, belly and tail slate. Secondaries when closed bay. Wing bar slate. Hackle and saddle feathers orange-red with blue centre stripe. Shoulders deep crimson-scarlet. Fluff dark slate.
Female plumage: Hackle golden striped. Breast and muff salmon. Body, wings and tail blue, finely peppered with golden brown. Fluff dark slate.

The pyle

Male plumage: The pyle is marked exactly like the black-red except that the black is exchanged for a clear cream-white. Secondaries bay.
Female plumage: Creamy-white with salmon breast and golden striped hackle.

The crele

Male and female plumage: Neck hackle straw barred with gold or black. Back and shoulder bright gold-chestnut barred with straw-yellow. Wing bar dark grey barred with pale grey; primaries and secondaries dark grey barred with pale; outer web of secondaries chestnut, the chestnut only showing when wing closed. Saddle hackle pale straw barred gold. Breast and underparts dark grey. Tail and tail coverts dark grey barred with light grey.

The spangled

Male and female plumage: These have white tips to their feathers. The more of these spots and the more regularly they are distributed the better. The male should show white ends to the feathers on hackle and saddle. The colour may be red, black or brown, or a mixture of all three. Fluff white.

The cuckoo

Male and female plumage: Light blue-grey ground colour, each feather crossed with broad bands of dark blue-grey. In the male, a lighter shade is permissible. Undercolour banded but of a lighter shade. Beak light horn or bluish. Legs and feet white with blue spots.

The black

Male and female plumage: Black with green sheen.

The white

Male and female plumage: Snow-white throughout.

In both sexes and all colours

Comb and face bright red. Eyes dark orange. Beak and nails horn. Legs in all colours except cuckoo, willow to olive or slate.

Weights

Male 2.70–3.20 kg (6–7 lb)
Female 2.25–2.70 kg (5–6 lb)

Scale of points

Type and carriage	20
Crest and muffling	25
Comb	10
Other head points	5
Feet and legs	5
Colour	20
Condition and handling	15
	100

Serious defects or disqualifications

Cut-away breast. Roach back. Wry or squirrel tail. Crest too small or too large, e.g. Poland type. Absence of crest or muffling. Comb other than of pea type. Comb lopped or twisted. Any comb other than minimal in female. Pearl eye. Feathered legs. Legs other than standard colour. Uneven or splashed breast colour. In males white base in tail. In lavenders any straw or brassy tinge.

BANTAM

The standard to be an exact miniature of the large fowl.

Weights

Male 740–850 g (26–30 oz)
Female 680–790 g (24–28 oz)

Serious defects

As large fowl, plus low wing carriage. High tail carriage. Any tendency to rose comb. Scale of points as in large fowl.

RUMPLESS ARAUCANA

LARGE FOWL

Origin: Chile
Classification: Light: Soft feather
Egg colour: Blue or green

The Rumpless Araucana also has its origins in South America. It was introduced to Europe by Professor S. Castello in the early 1920s. The ear-tufts of feathers are unique to the breed in that they grow from a fleshy pad adjacent to the ear-lobe. Rumpless Araucanas lay a large egg in relation to body size and are as productive as the tailed Araucanas.

General characteristics: male

Carriage: Alert, active and assured.
Type: Body moderate in length, broad at shoulders. Back flat and slightly sloped. Rump well rounded with saddle feathers flowing over stern. Breast full, round and deep. Wings medium in length, carried close to the body and well up. Saddle hackle well developed. Tail entirely absent, with no uropygium (parson's nose).
Head: Moderately small. Beak medium stout, curved. Eyes bold and expressive. Comb small pea. Face moderate muffling. Ear-lobes small and concealed by ear-tufts. These originate from a gristly appendage arising from behind and just below the ear hole. The tufts of feathers, numbering from 5 to 15, grow from this pad. The tufts should be of a

good length, matching in size and extending from the ears backwards in a well-defined sweep, or project horizontally. Wattles very small.
Neck: Medium length, well furnished with hackle feathers.
Legs and feet: Medium in length, straight and well set apart. Toes, four, strong and well spread.

Female

The general characteristics are similar to those of the male allowing for the natural sexual differences.

Colour

Male and female plumage: As for Araucanas.

In both sexes and all colours

Eyes dark orange. Legs and feet olive or slate.

Weights

Excess weight to be penalized.
Male 2.70 kg (6 lb)
Female 2.25 kg (5lb)

Scale of points

Type and carriage	20
Ear-tufts	25
Comb	5
Other head points	5
Feet and legs	5
Colour	15
Condition and handling	25
	100

Serious defects

Non-standard comb. Unmatched ear-tufts. Shape other than standard, e.g. narrow body. Any tail feathers (incomplete rumpless). Fluff showing below saddle hackle.

Disqualifications

No ear-tufts, single ear-tuft, crest. Uropygium (parson's nose).

BANTAM

These should be a true miniature of the large Rumpless Araucana. As the large Rumpless fowl is historically and naturally a small breed, it follows that great care must be taken to keep the bantams within the approved weight limits. Colours at present include black-red, black and white.

Weights

Excess weight to be penalized.
Male 910 g (32 oz)
Female 790 g (28 oz)

ASIL

LARGE FOWL

Origin: Asia
Classification: Rare: Hard feather
Egg colour: Tinted

The Asil is probably the oldest known breed of game fowl, having been bred in India for its fighting qualities for over 2000 years. The name Asil is derived from the Arabic and means 'of long pedigree'. In its native land the Asil was bred to fight, not with spurs, but rather with its natural spurs covered with tape and the fight became a trial of strength and endurance. Such was the fitness, durability and gameness of the contestants that individual battles could last for days. This style of fighting produced a powerful and muscular bird with a strong beak, thick, muscular neck and powerful legs and thighs together with a pugnacious temperament and a stubborn refusal to accept defeat.

Never very numerous in Britain, the Asil has nevertheless always attracted a few dedicated admirers prepared to cope with its inborn desire to fight, a characteristic shared by the females who are poor layers but extremely good mothers.

General characteristics: male

Carriage: Upright, standing firmly and well on his legs. Sprightly and quick in movement. When seen in profile the eye should be directly above the middle toenail.
Type: Chest wide and well thrown out. Hard and muscular and feeling remarkably flat in the hand. Back broad and flat tapering to a fairly narrow stern but very well developed and strong at the root of the tail. Viewed from above the body should appear to be heart-shaped. Wings carried well out from the body at the shoulders and should be muscular where they join the body but otherwise carrying very little flesh and covered with hard feathers and tough, rather short quills. Sickle feathers narrow and scimitar-shaped, drooping from the base. Saddle feathers pointing backwards more so than in other breeds.
Head: Skull broad with large and square jaw bones and large cheek bones covered with tough leathery skin. Beak short, thick, powerful, shutting tight. Eyes bright and bold set in oval pointed eyelids. Iris pearl colour but occasionally seen slightly bloodshot or with a yellowish tinge. Comb triple or pea, very hard fleshed and set low. No wattles.
Neck: Medium length carried slightly curved to give a short appearance. Thick and very hard to the touch and covered with short, hard and wiry feathers. Throat clean cut with bare skin extending well down the neck.
Legs and feet: Thick and square with a noticeable indentation down the front of the leg where the scales meet. Toes, four, straight, strong and tapering with broad, curved toenails. Thighs not too long, round, hard muscular and when viewed from the front should be in line with the body and not the shoulders.
Plumage: Short and wiry. Difficult to break and with little or no underfluff. Patches of bare skin showing red are to be seen on the breastbone, wing joints and thighs.
Handling: Firm and muscular. Heavier in the hand than appearance would at first suggest.

Female

The general characteristics are similar to those of the male allowing for the natural sexual differences.

Colour

Male and female plumage: There are no fixed colours, the principal colours seen today being light red and dark red with grouse-coloured and red-wheaten females. Greys, spangles, blacks, whites, duckwings and piles have been seen.

Red Asil male, large

In both sexes and all colours

Except that the comb, face, jaw and throat are red, there is no fixed colour for beak or legs although the beak is generally seen in yellow or ivory colour and the legs willow, white or dark olive.

Weights

Male 1.80–2.70 kg (4–6 lb)
Female 1.35–2.25 kg (3–5 lb)

Scale of points

Head (skull and beak 10, eyes 5, comb 5)	20
Neck	10
Wings	5
Thighs, shanks and feet	15
Body shape and stern	15
Plumage	10
Carriage	15
Condition	10
	100

Serious defects

Any evidence of alien blood, e.g. red or dark eyes, red markings on the side of the shanks, etc. Round shanks. Duck feet. High tail carriage. Wry tail. Roach back. Stork legged or in-kneed. Any other deformity.

BANTAM

Bantams should be exact miniatures of their large fowl counterparts and so standard, colour and scale of points apply.

Weights

Male 1130 g (40 oz)
Female 910 g (32 oz)

AUGSBURGER

LARGE FOWL

Origin: Germany
Classification: Light: Rare
Egg colour: White

This breed was developed from crosses with La Flèche and other breeds beginning in 1870 at Augsburg and the Black Forest area. The La Flèche gave the duplex gene for the distinctive cup comb of the Augsburger. They have a strong following among German fanciers who appreciate both their attractive appearance and practical qualities of egg and meat production.

General characteristics: male

Carriage: Somewhat upright.
Type: Body medium sized, powerful and carried slightly sloping. Back broad, rather long, sloping moderately downwards. Breast full and broad, carried moderately high. Tail long, well furnished with broad, long sickles and side hangers. Main tail fanned and carried well out. Wings neatly tucked under saddle hackle.
Head: Medium sized, rather broad, face smooth. Ear-lobes smooth and oval in shape. Wattles medium sized and rather thin. Comb of cup type: after one or two single serrations at the front, the comb divides to form a cup, as upright as possible and appearing to be closed at the rear. The comb should be well serrated, and without excrescences in the centre. Eyes bright and prominent.
Neck: Of medium length, well feathered.
Legs and feet: Legs of medium length. Shanks and toes (four) finely scaled with no trace of feathering.
Plumage: Fairly full and very glossy.

Female

With the exception of the tail, which is carried a little lower than the male, the general characteristics are the same, allowing for the natural sexual differences.

Colour

Male and female plumage: Glossy green-black throughout.
In both sexes: Comb, face and wattles bright red. Ear-lobes pure white. Eyes dark. Beak dark. Shanks and feet slate-grey.

Weights

Male 2.30–3.00 kg (5–6½ lb)
Female 2.00–2.50 kg (4½–5½ lb)

Scale of points

Type and carriage	20
Head	35
Plumage	20
Legs and feet	10
Condition	15
	100

Serious defects

Any body deformity. Plumage showing white or other than green sheen. Serious faults in comb structure.

BANTAM

Augsburger bantams follow the large standard.

Weights

Male	900 g (32 oz)
Female	800 g (28 oz)

AUSTRALORP

LARGE FOWL

Origin: Great Britain
Classification: Heavy: Soft feather
Egg colour: Tinted to brown

The claim that the Australorp – an abbreviation of Australian black Orpington – is the prototype of the black Orpington, as originally made by Mr W. Cook, has never been questioned. Its breeders emphasized that its true utility type gives to poultrymen the Orpington at its best, an excellent layer and a good table fowl, with white skin.

It was around 1921 that large importations of stock birds were made from Australia into this country and an Austral Orpington Club founded. Later the breed name of Australorp was adopted, and this remains today.

General characteristics: male

Carriage: Erect and graceful, denoting an active fowl, the head being carried well above the tail line.
Type: Body deep and broad, showing somewhat greater length than depth. Back broad across shoulders and the saddle, with a sweeping curve from neck to tail. Breast full and rounded, carried well forward without bulging; breastbone long and straight. Wings compact and carried closely in, the ends being covered by the saddle hackles. Tail full and compact, rising gradually from the saddle in an unbroken line; the sickles gracefully curved, but not long and streaming.
Head: Finely modelled with skull rounded. Beak slightly curved, strong, of medium length. Eyes, large, prominent and expressive; high in skull standing out well when viewed

Australorp male, large

Australorp female, bantam

from front or back. Comb single, medium in size, erect, evenly serrated (four to six serrations) and blade tending downwards without touching the neck, texture fine, but not of polished appearance. Face full, fine in texture, clean, free from feathers, wrinkles and overhanging brows. Ear-lobes small and elongated. Wattles medium in size, rounded at bottom and corresponding in texture to comb.
Neck: Fairly long, fine at the junction of head, with a gradual outward curve to the back, widening distinctly at the shoulders.
Legs and feet: Legs medium in length, strong, rounded in front and spaced well apart, the hocks nearly covered by body feathering, and the whole of the shanks showing below the underline. Shanks and feet (four toes) free from feathers or down.
Plumage: Feathering soft but close, with a minimum of fluff, the lower body fluff only sufficient to cover the thighs.
Skin: Fine in texture.

Female

The general characteristics are similar to those of the male, allowing for the natural sexual differences. The pelvic bones should be pliable, not showing an excess of fat or gristle; the abdominal skin being pliable without an excess of internal fat. All these parts to be of fine texture; any indication of coarseness should be discountenanced.

Colour

The black

Male and female plumage: Black with lustrous green sheen.

In both sexes: Beak black. Eyes black or dark brown iris, black preferred. Face, comb, ear-lobes and wattles bright red. Legs and feet black with white soles. Skin white.

The blue

Male plumage: Hackles, saddle, wing bow, back and tail a uniform dark slate-blue. Remainder medium slate-blue, each feather to show a wide band of lacing of a darker shade.
Female plumage: Head and neck dark slate-blue, remainder medium slate-blue, laced with a darker shade.

Undercolour to tone with surface colour in both sexes.

In both sexes: Beak blue or black, black preferred. Eyes black or very dark brown, black preferred. Comb, face, wattles and ear-lobes bright red. Legs and feet black or blue. Toenails preferably white. Skin white. Soles of feet white.

Weights

Cock	3.85–4.55 kg (8½–10 lb);	cockerel	3.40–4.10 kg (7½–9 lb)
Hen	2.95–3.60 kg (6½–8 lb);	pullet	2.50–3.20 kg (5½–7 lb)

Scale of points

Type	35
Head (eyes 10, face 5, skull 5, comb and wattles 5)	25
Plumage (colour, quality and character of feathering)	12
Texture and freedom from coarseness	15
Condition	8
Legs and feet	5
	100

Serious defects

Red, yellow or white in feathers. Permanent white in ear-lobes.

Defects (for which birds should be passed)

Any deformity such as wry tail, roach back, crooked breastbone, crooked toes, webbed feet. Yellow or willow colour in legs or feet. Yellow or pearl coloured eyes. Feathering on shanks or feet. Side sprigs on comb. Split or twisted wing and slipped wing.

BANTAM

Australorp bantams are to be exact miniatures of their large fowl counterparts and so standard, colour and scale of points to apply.

Weights

Male 1020 g (36 oz) max.
Female 790 g (28 oz) max.

AUTOSEXING BREEDS

An autosexing breed is one in which the chicks at hatching can be sexed by their down colouring. It was when crossing the gold Campine with the barred Rock in 1929 that Professor R. C. Punnett and Mr M. S. Pease discovered the basic principle in their experimental work at Cambridge, and made the Cambar.

Barring is sex-linked, there being a double dose in the male and a single dose in the female, the barring being indicated by the light patch on the head of the chick. This light patch is very similar in chicks of both sexes having black down, but when the barring is transferred to a brown down there is a marked difference. The light head-spot on the female chick (one dose) is small and defined, while on the male chick (double dose) it spreads over the body. For that reason, the down colouring in the day-old cockerel is much paler, and the pattern of markings more blurred, than in the newly hatched pullet chick, which has the sharper pattern of markings.

Standards which have been passed by the Poultry Club are gold and silver Brussbar; Brockbar; gold, silver and cream Legbar; gold and silver Cambar; gold and silver Dorbar; Rhodebar; silver Welbar; Wybar. The cream Legbar, Legbar, Rhodebar, Welbar and Wybar standards are given below and all other standards for autosexing breeds are held by the Rare Poultry Society.

LEGBAR

LARGE FOWL

Origin: British
Classification: Light: Rare
Egg colour: White or cream. *Cream Legbar:* blue, green or olive.

The Legbar comes in three varieties, gold, silver and cream. The cream variety is a crested breed which lays a blue, green or olive egg. Thereafter, the characteristics are the same for all three colours. The gold and silver Legbars are barred Leghorns whereas the cream Legbar has had a dose of Araucana blood to give it its crest and egg colour.

General characteristics: male

Carriage: Very sprightly and alert, with no suggestion of stiltiness.
Type: Body wedge shaped, wide at the shoulders and narrowing slightly to root of tail. Back long, flat and sloping slightly to the tail. Breast prominent, and breastbone straight. Wings large, carried tightly and well tucked up. Tail moderately full at an angle of 45° from the line of the back.
Head: Fine. Cream variety: crest small, compact and carried well back from eyes, falling off the back of the head below the extended comb. Beak stout, point clear of the front of the comb. Eyes prominent. Comb single, perfectly straight and erect, large but not overgrown, deeply and evenly serrated (five to seven spikes broad at the base), extending well beyond back of the head and following, without touching, the line of the head, free from 'thumb marks' or side sprigs. Face smooth. Ear-lobes well developed, pendant, smooth and free from folds, equally matched in size and shape. Wattles long and thin.
Neck: Long and profusely covered with feathers.
Legs and feet: Legs moderately long. Shanks strong, round and free of feathers. Flat shins objectionable. Toes four, long, straight and well spread.
Plumage: Of silky texture, free from coarse or excessive feather.
Handling: Firm, with abundance of muscle.

Female

The general characteristics are similar to those of the male, allowing for the natural sexual differences, except that the comb may be erect or falling gracefully over either side of the face without obstructing the eyesight, and the tail should be carried closely and not at such a high angle. Cream variety: the crest of the female is somewhat fuller and larger than the male, but does not obstruct the eyes.

Colour

The gold

Male plumage: Neck hackle pale straw, sparsely barred with gold and black. Back, shoulder coverts and wing bow pale straw barred with bright gold-brown. Wing coverts (or wing bar) dark grey barred; primaries and secondaries dark grey barred, intermixed with white, upper web of secondaries also intermixed with chestnut. Saddle hackle pale straw barred with bright gold-brown, as far as possible without black. Breast, underparts, dark grey barred. Tail grey barred; sickles paler. Tail coverts grey barred.
Female plumage: Hackle pale gold, marked with black bars. Breast salmon, clearly defined. Body dark smoky or slaty grey-brown with indistinct broad soft barrings, the individual feather showing paler shaft and slightly paler edging. Wings dark grey-brown. Tail dark grey-black with slight indication of lighter broad bars.

The silver

Male plumage: Neck hackle silver, sparsely barred with dark grey but tips of feathers fade off to pure silver. Saddle hackle silver, barred with dark grey, the feathers tipped with silver. Back and shoulder coverts silver, with dark grey barring, the feathers tipped with silver. Wing bow dark grey with silver grey barring; primaries dark grey, some white permissible; secondaries dark grey with tips of upper web white. Breast evenly barred dark grey and silver-grey, with well defined outline. Tail and tail coverts evenly barred dark grey and silver-grey, sickles being paler.
Female plumage: Head and neck hackle silver, with black striping, softly barred grey. Breast salmon, clearly defined. Body silver-grey, with indistinct broad soft barring, individual feathers showing lighter shaft and edging. Wings silver-grey, as free from

chestnut as possible; primaries silver-grey, as free from white as possible; secondaries silver-grey, upper web a lighter grey mottled. Tail silver-grey with indistinct soft barring.

In both sexes: Beak yellow or horn. Eyes orange or red, pupils clearly defined. Comb, face and wattles bright red. Ear-lobes pure opaque white (resembling white kid) or cream, the former preferred. Slight pink markings and pink edging does not unduly handicap an otherwise good bird for utility purposes. Legs and feet yellow, orange or light willow in the female.

The cream

Male plumage: Neck hackles cream, sparsely barred. Saddle hackles cream barred with dark grey, tipped with cream. Back and shoulders cream with dark grey barring, some chestnut permissible. Wings, primaries dark grey, faintly barred, some white permissible; secondaries dark grey more clearly marked; coverts grey barred, tips cream, some chestnut smudges permissible. Breast evenly barred dark grey, well defined outline. Tail evenly barred grey, sickles being paler, some white feather permissible. Crest cream and grey, some chestnut permissible.

Female plumage: Neck hackles cream, softly barred grey. Breast salmon, well defined in outline. Body silver-grey, with rather indistinct broad soft barring. Wings, primaries grey peppered; secondaries very faintly barred; coverts silver-grey. Tail silver-grey, faintly barred. Crest cream and grey, some chestnut permissible.

In both sexes: Beak yellow. Eyes orange or red. Comb, face, and wattles red. Ear-lobes pure opaque, white or cream, slight pink markings not unduly to handicap an otherwise good male. Legs and feet yellow.

Weights

Male	2.70–3.40 kg (6–7½ lb)
Female	2–2.70 kg (4½–6 lb)

Scale of points

Type	30
Colour	20
Head	20
Legs	10
Condition	10
Weight	10
	100

Serious defects

Male's comb twisted or falling over. Ear-lobes wholly red. Any white in face. Legs other than orange, yellow or light willow. Squirrel tail.

Disqualifications

Side sprigs on comb. Eye pupil other than round and clearly defined. Crooked breast. Wry tail. Any bodily deformity.

The gold: down colour

Female: Brown stripe type. The stripe should be broad and very dark brown, extending over the head, neck and rump. The edges of the stripe should be clearly defined rather than blurred and blending with the ground colour – the sharper the contrast, especially over the rump, the better. The ground colour should be dark brown, though distinctly paler than the stripe. A pale ground colour and a narrow or discontinuous stripe are to be avoided. A light head spot should be visible, though usually it is small. It should be well defined in outline and should show up as clearly as possible against the brown background.

Male: The down is much paler in shade, the pattern being blurred and washed out from head to rump.

The silver: down colour

Female: Silver-grey type. The stripe should be very dark brown, extending over the head, neck and rump. The edges of the stripe should be clearly defined, not blurred and blending with the ground colour – the sharper the contrast, especially over the rump, the better. The stripe should be broad; a narrow or discontinuous stripe should be avoided. A light head patch should be visible, clearly defined in outline, showing up brightly against the dark background.
Male: The down is much paler in tint, the pattern being blurred and washed out from head to rump; it may best be described as pale silvery-slate.

The cream: down colour

As silver.

BANTAM

Bantam Legbars should follow the exact standard for the large fowl.

Weights

Male 850 g (30 oz)
Female 620 g (22 oz)

RHODEBAR

LARGE FOWL

Origin: British
Classification: Heavy: Rare
Egg colour: Brown

General characteristics: male

Carriage: Upright and graceful.
Type: Body large, fairly deep, broad and long. Back broad, long and somewhat horizontal in outline. Breast broad, full and well rounded. Wings carried well up, the bows and tips covered by the breast feathers and saddle hackle. Tail rather small, rising slightly from the saddle, the sickle of medium length, well spread and nicely curved, the coverts sufficiently abundant to cover the stiff feathers.
Head: Strong, but not thick. Beak moderately curved, short and stout. Eyes large and bright. Comb single, medium size, straight, upright, well set on, with well-defined serrations, and free from side sprigs. Face smooth. Ear-lobes of fine texture, well developed and pendant. Wattles to correspond with size of comb and moderately rounded.
Neck: Of medium length and profusely covered with feathers flowing over the shoulders, but not too loosely carried.
Legs and feet: Legs wide apart and of medium length, stout and strong and free from feathers. Thighs large with well-rounded shanks of medium length. Toes, four, strong, straight and well spread.
Handling: Firm, with abundance of muscle.
Plumage: Of silky texture, free from coarse or excessive feather.

Female

The general characteristics are similar to those of the male, allowing for the natural sexual differences.

Colour

Male plumage: Hackle deep red-gold barred, with centres black and grey-white barred, the black centre portions rather longer than the grey-white; the front of the cape showing less black, the feathers towards the tips of the cape lying on the back showing wider black and grey-white barring. Wing primaries, lower web red-gold, faintly barred, upper grey and white barred, slightly gold tinted; secondaries, the whole alternately black, white and gold barred, lower web showing more gold; flight coverts very bright red-gold and white barred; tips red-gold. Wings bows very brilliant chestnut red and gold barred. Tail, including sickles, uniform black and white barring from tip to base, including the shaft; tips black. Saddle hackle deep red-gold and grey-white and narrower black barring towards the tips. Back and saddle deep red-gold barred, with occasional black bars towards the end of the feathers. Undercolour light creamy buff. Breast uniformly barred, deep red-gold and creamy white and black.
Female plumage: Hackle deep buff-red with bright chestnut edges, each feather with deep buff, gold, black and white narrow barring, the barring becoming narrower as it approaches the lower cape feathers. Tail feathers black with reddish tinge. Wing primaries upper web red-buff, lower black; secondaries buff-red. Remainder, general surface dark buff-red barred with buff and buff-red, the tips of the feathers of the lighter colour. Undercolour creamy buff-red, as deep as possible. Quills yellow.

In both sexes: Beak red-horn or yellow. Eyes orange or red, pupils clearly defined. Comb, face, ear-lobes and wattles bright red. Legs and feet bright yellow.

Weights

Cock	3.85 kg ($8\frac{1}{2}$ lb) minimum;	cockerel	3.60 kg (8 lb)
Hen	2.90 kg ($6\frac{1}{2}$ lb) minimum;	pullet	2.50 kg ($5\frac{1}{2}$ lb)

Scale of points

Type	30
Colour	20
Legs	10
Condition	15
Head	20
Weight	5
	100

Serious defects

Male's comb twisted or falling over. Ear-lobes other than red. Legs other than yellow, orange or light willow. Squirrel or wry tail. Side sprigs on the comb. Eye pupils other than round and clearly defined. Crooked breast or any bodily deformity.

BANTAM

Bantam Rhodebars should follow the exact standard for the large fowl.

Weights

Male	1020 g (36 oz)
Female	790 g (28 oz)

WELBAR

LARGE FOWL

Origin: British
Classification: Light: Rare
Egg colour: Brown

General characteristics: male

Carriage: Upright, alert and active.
Type: Body well built on good constitutional lines. Back broad and long. Breast full, well rounded and broad. Wings moderately long, carried close to side. Tail fairly large and full, carried high, but not squirrel. Abdomen long, deep and wide.
Head: Refined. Beak strong, short and deep. Eyes large, bright. Comb single, medium size, firm and upright, free from any twists or excess, clear of the nostrils, fine texture, five to seven broad and even serrations, the back following closely, but not touching line of the skull and neck. Face smooth and without overhanging eyebrows. Ear-lobes small and almond shaped. Wattles of medium size, fine texture, close together.
Neck: Fairly long, slender at top, finishing with abundant hackle.
Legs and feet: Thighs to show clear of the body. Shanks of medium length and bone, well set apart, free from feathers with soft sinews and free from coarseness. Toes, four, long, straight and well spread out.
Plumage: Tight, silky, free from excess or coarseness and free from bagginess at the thighs.
Handling: Compact, firm and neat in bone throughout.

Female

The general characteristics are similar to those of the male, allowing for the natural sexual differences.

Colour

The silver

Male plumage: Head silver. Hackles silver, black ticking permissible. Back, shoulders, coverts and wing bow silver. Wings coverts (or bar) black barred; primaries and secondaries, inner web black barred, outer web silver. Tail black barred. Breast black barred with silver mottling.
Female plumage: Head and hackle silver with black striping barred with white. Breast salmon. Back and wing bow and bar, light grey, faintly barred, free from salmon smudges. Primaries and secondaries, outer web silver, coarsely stippled with dark grey, inner web dark and faintly barred. Tail dark grey and faintly barred.

The gold

Male plumage: Head gold. Hackles gold, black ticking permissible. Back, shoulder coverts and wing bow gold. Wing coverts (or bar) black barred; primaries and secondaries, inner web black barred, outer web gold. Tail black barred. Breast black barred with gold mottling.
Female plumage: Head and hackle gold with black striping barred with white. Breast rich salmon. Back and wing bow and bar mid-grey, faintly barred, free from salmon smudges. Primaries and secondaries, outer web gold coarsely stippled with dark browny grey, inner web brown-grey and faintly barred. Tail brown-grey and faintly barred.

In both sexes and colours

Beak yellow, eyes red. Comb, face, ear-lobes and wattles bright red. Legs and feet yellow.

Weights

Male 2.95–3.40 kg ($6\frac{1}{2}$–$7\frac{1}{2}$ lb)
Female 2.25–2.70 kg (5–6 lb)

Scale of points

Type	30
Colour	20
Head	20
Legs	10
Condition	10
Weight	10
	100

Serious defects

Side sprigs to the comb. White in lobe. Feathers on the legs, hocks or between the toes. Comb other than single. Other than four toes. Gold feathers on the silver male. Legs other than yellow. Badly crooked or duck toes. Any bodily deformity. Coarseness, beefiness and anything that interferes with the productiveness and the general utility of the breed.

BANTAM

Bantam Welbars should follow exactly the standard for the large fowl.

Weights

Male 900 g (32 oz)
Female 745 g (26 oz)

WYBAR

LARGE FOWL

Origin: Great Britain
Classification: Heavy: Rare
Egg colour: Tinted

Like all other autosexing breeds, the Wybar was first launched at the Cambridge School of Agriculture in 1941. Breeds used in its make-up were the light Sussex, Brussbar (barred brown Sussex), Canadian barred Rocks and, later on, Rhode Island Reds. The main Wyandotte variety used was the silver laced. The outcome was a large, all-round bird suitable for both eggs and meat.

General characteristics: male

Type: Body short and deep, well rounded at sides. Back short, with broad full saddle to tail with concave sweep. Breast full round, with a straight keel. Wings medium size closely folded to the side. Tail full but short, well developed and spread at base, the true tail feathers carried rather upright, sickles medium length and gracefully curled.
Head: Broad and short, beak stout and curved. Comb rose firm, square and low in front, tapering evenly towards the back and ending in a well-defined leader following the curve of the neck. Top of comb evenly covered with small rounded joints. Face smooth and fine. Ear-lobes oblong. Wattles medium, fine and rounded.
Neck: Well arched, medium length, with full hackle.
Legs and feet: Legs of medium length, thighs well covered with soft feathers, the fluff

Barnevelder male, large

Barnevelder female, bantam

The partridge

Male plumage: Neck and saddle hackles red-brown with distinct but small black tip; fluff grey; quill red-brown. Breast black (beetle-green). Abdomen and thighs black (beetle-green) with black down. Back, cape and wing bow red-brown with wide black tip; fluff grey; quill red-brown. Wing bar black; bay brown; primaries, inner edge black, outer brown; secondaries, inner edge black, outer brown (seen as wing is closed). Tail, main feathers black with beetle-green sheen; coverts, upper black, lower red-brown peppered with black; sickles black with beetle-green sheen. All visible black feathers with beetle-green sheen.

Female plumage: Hackle black with beetle-green sheen. Breast, saddle, back and thighs red-brown ground evenly stippled with small black peppering, clear of defined inner lacing or pencilling, each feather with glossy black outer lacing, not so broad as to make the bird appear black when seen in the show-pen. Wing primaries, inner edge black, outer brown peppered with black; secondaries, outer edge brown evenly stippled with small black peppering. Tail, main feathers black, coverts peppered. Undercolour grey.

The silver

Male plumage: Hackle silver with black centres. Breast silver with black edging. Back and saddle black centre with white edges. Undercolour silver-grey. Wing primaries black; secondaries black edged with white. Tail black with beetle-green sheen; sickles edged with white.

Female plumage: Hackle black centre with white edges, a little rust permissible. Breast white, slightly peppered, outside edge black. Wing primaries black inside, white outside, slightly peppered; secondaries well peppered.

In both sexes and all colours

Beak yellow with dark point (in the silver, horn) Eyes orange. Comb, face, wattles and ear-lobes red. Legs and feet yellow.

Weights

Cock	3.20–3.60 kg (7–8 lb);	cockerel	2.70–3.20 kg (6–7 lb)
Hen	2.70–3.20 kg (6–7 lb);	pullet	2.25–2.70 kg (5–6 lb)

Scale of points

Type and size	30
Colour	25
Texture	15
Head	10
Legs and feet	10
Health and condition	10
	100

Minor defects

White in undercolour, flights, tails, wings, sickles or fluff.

Serious defects

White in lobes. Squirrel or wry tail. Feathered legs or toes. Side sprigs on comb. Crooked toes. High or roached back. Seriously deformed breastbones. More than four toes on either foot. Black legs.

BANTAM

Barnevelder bantams are exact replicas of their large fowl counterparts and so standard, defects and scale of points apply. However, silvers are not standardized in bantams.

Weights

Male 910 g (32 oz) max.
Female 740 g (26 oz) max.

BELGIAN BEARDED BANTAMS

The only varieties of Belgian bantams yet standardized in Britain are Barbu d'Anvers (Bearded Antwerp), Barbu d'Uccle (Bearded Uccle) and Barbu de Watermael (Bearded Watermael).

Belgian Bearded bantams are old-established True Bantams, without counterparts in large breeds. Each of the two varieties (d'Anvers and d' Uccle) has many colour variations, some of them intricate and all attractive.

BARBU D'ANVERS

General characteristics: male

(The Barbu d'Anvers is always rose combed and clean legged.)
Carriage and appearance: Small, proud, standing bold upright, with the head thrown well back; proud and provoking (appearing always ready to crow) with characteristic great development of neck hackle.
Type: Body broad and short, with arched breast carried well up. Back very short, slanting downwards to tail. Wings medium length, carried sloping towards ground. Tail carried almost perpendicularly, the main tail feathers strong and not hidden by the narrow sickle feathers; the two largest sickles slightly curved and sword-shaped, the remainder in fan-like tiers to junction with saddle hackle.
Head: Appearing rather large. Beak short, strong and curved, carrying a longitudinal band of light or dark colour in keeping with the plumage. Comb curved broad in front, ending in a leader or spike at rear; for preference covered with small tooth-like points, or alternatively hollowed and ridged. Point or leader to follow line of neck. Eyes large and prominent, as dark as possible, colour to vary in keeping with plumage. Face covered with relatively long feathers, standing away from the head, sloping backwards and forming whiskers which cover ears and ear-lobes. Brow heavily furnished with feathers. Beard composed of feathers turned horizontally backwards from both sides of the beak and from the centre vertically downwards, the whole forming a trilobe and giving a muffed effect. Ear-lobes small, wattles rudimentary only, but preferably none.
Neck: Of moderate length, the hackles thick and convexly arched (boule), entirely covering back and base of neck forming closely joined cape at front.
Legs and feet: Thighs short, with medium-length shanks free from feathers. Toes, four, strong and straight, with nails of same colour as the beak.

Female

With certain exceptions the general characteristics are similar to those of the male, allowing for the natural sexual differences.
Carriage and appearance: A little bird, compact, plump, very lively, with characteristically full, rounded neck hackle and well-developed whiskers.
Head: Appearing broader than that of the male and more owl-like.
Neck: Hackle inclining backwards and forming a ruffle behind the neck, with feathers broader and more developed than in the male. The female hackle, contrary to that of the male, diminishes in thickness towards bottom of neck.
Tail: Short, carried sloping upwards, slightly curved towards the end and a little open.

Quail Barbu d'Anvers male

Quail Barbu d'Anvers female

Weights

As small as possible. The British Belgian Bantam Club does not advocate a weight standard for the breed but, purely as a general guide, suggests with usual variations for age and maturity the following:
Male 680–790 g (24–28 oz) max.
Female 570–680 g (20–24 oz) max.

Scale of points

Type (muff and beard 15, neck hackle 15, wings and tail 10)	40
Head, comb and beak	10
Colour	15
Size	15
Legs and feet	5
General appearance	15
	100

Serious defects

Wattles strongly developed. Conspicuous ear-lobes. Squirrel or wry tail. Excessive length of leg.

Disqualifications

Any trace of faking. Wattles cut or removed. Single comb. Absence of beard or whiskers. Feathers on shanks or feet. More than four toes. Yellow colouring of legs, feet or skin.

BARBU D'UCCLE

General characteristics: male

(The Barbu d'Uccle is always single combed and feather legged.)
Carriage and appearance: Typically male with a majestic manner, short and broad, with characteristic heavy development of plumage.
Type: Body broad and deep. Back very broad, almost hidden by enormous neck hackle. Breast extremely broad, the upper part very developed and carried forward, the lower part resembling a breast plate. Wings close, fitting tight to body, sloping downwards and incurved towards but not beyond the abdomen; wing butts covered by neck hackle and tips (or ends of flights) covered by saddle hackle, which should be abundant and long. Tail well furnished, close and carried almost perpendicular to line of back, the two main sickles slightly curved, the remainder in regular tiers and fan-like down to junction with saddle hackle.
Head: Slender and small, with a longitudinal depression towards the neck. Beak short and slightly curved. Comb single, fine, upright, less than average size, evenly serrated, rounded in outline, blade following line of neck. Eyes round, surrounded with bare skin. Brow heavily covered with feathers becoming gradually longer towards the rear, with a tendency to join behind the neck. Beard as full and developed as possible, composed of long feathers turned horizontally from the two sides of beak, and vertically under the beak downwards, the whole forming three ovals in a triangular group. Ear-lobes inconspicuous. Wattles as small as possible.
Neck: Furnished with silky feathers starting behind the beard at sides of throat, with a tendency to join behind the neck to form a mane; hackle very thick and convexly arched (boule), reaching to shoulder and saddle and covering the whole back.
Legs and feet: Legs strong and well apart, the hocks having clusters of long, stiff feathers close together, starting from the lower outer thigh, inclined downwards and following

Porcelaine Barbu d'Uccle male

Millefleur Barbu d'Uccle female

outline of wings. Front and outside of shanks must be covered with feathers, short at top of shanks and gradually increasing in length towards the foot feather; footings turned outwards horizontally, with ends slightly curved backwards. Outer toe and outside of middle toe covered with feathers similar to shank feather.

Female

With certain exceptions the general characteristics are similar to those of the male, allowing for the natural sexual differences.
Carriage and appearance: A quiet little bird, short, thick and cobby.
Beard: Resembling that of the male but formed with softer and more open feathers.
Neck: Hackles very thick and convexly arched, composed of broad and rounded feathers, the shape of the mane resembling that of the male.
Tail: Short, flat in width and not high, the lower main feathers diminishing evenly in length.

Weights

Dwarf, as small as possible. The British Belgian Bantam Club does not advocate a weight standard but, purely as a general guide, suggests, with usual variations for age and maturity:

Male	790–910 g (28–32 oz) max.
Female	680–790 g (24–28 oz) max.

Scale of points

Type (muff and beard 15, neck hackle 15, feet and hocks 15, wings and tail 10)	55
Head, comb and beak	10
Colour	15
Size	5
General appearance	15
	100

Serious defects

Strongly developed wattles. Conspicuous ear-lobes. Squirrel or wry tail. Excessive length of leg.

Disqualifications

Any trace of faking. Wattles cut or removed. Comb other than single. Absence of beard or whiskers. Poorly feathered shanks or feet. More than four toes. Yellow legs, feet or skin.

BARBU DE WATERMAEL

General characteristics: male

(The Barbu de Watermael is always crested and clean legged.)
Carriage and appearance: Proud little bantam characterized by its beard and small crest. Always on the move, perky.
Type: Breast rounded, carried forward and well up. Back short and sloping backwards. Wings medium length, carried sloping towards ground, curving beneath tail in female, spread lower by the male. Tail slightly open and carried well off the perpendicular. The sickles quite short and only slightly curved.
Head: Appears large because of the crest and whiskers, skull normal. Rose comb, medium sized (length 3 cm, width 1 cm) covered with small tooth-like points, ending with three small leaders. Crest quite bushy, not too long, slightly erect and 'flying' backwards. Beak rather short, slightly curved. Ear-lobes and wattles rudimentary only and covered by

Blue Barbu de Watermael female

muff and beard. Ear-lobes preferably white. Muff and beard well developed and forming a trilobe.
Neck: Hackles thick forming a mane at the back.
Legs and feet: Thighs hidden by feathers of abdomen, shanks medium length (5 cm), fine. Toes four, smooth.

Female

Allowing for the natural sexual differences, the characteristics are the same as for the male except for the following.
Crest: Semi-globular, but much smaller than that of the Poland and not interfering with the sight of the bird.
Neck: Hackles not as thick as the male's but still forming a mane.
Back: A little longer than that of the male.
Tail: Closed, but carried at the same angle as that of the male.

Weights

Male	600–700 g (21–24$\frac{1}{2}$ oz) according to age
Female	450–550 g (15$\frac{1}{2}$–19$\frac{1}{2}$ oz) according to age

Scale of points

Type	50
Head and crest	10
Beard and hackle	10
Colour	10
Legs and feet	5
Wings	15
	100

Minor defects

Two leaders at end of comb instead of three. Gypsy face, or traces of dark pigmentation. Red ear-lobes.

Serious defects

Size too large. Light eyes. Wattles too developed. Beard insufficient or not trilobed. Tail open in hen or fan-shape in male, long sickles in male. Longitudinal furrow in top of the comb.

Disqualifications

Poland-type skull or crest, crest too narrow. Yellow colouring of legs, feet or skin.

Colours for all Belgian bantams

Main colours only are fully described. Belgian bantams exist in an extraordinary choice of colours, probably unequalled in any other breed, and much too numerous to be given in detail. The Barbu de Watermael is standardized only in black, blue, blue quail and quail at present.

Comb, ear-lobes and rudimentary wattles are red in all colour varieties.

The millefleur

Male plumage: This is a very intricate and attractive colour scheme. Briefly, the head is orange-red with white points. The beard is of black feathers laced with very light chamois, each feather ending with a round black spot with a white triangular tip. Neck hackle black with golden shafts, and broadly bordered with orange-red, each feather having a black end tipped with a white point. The extraordinary abundance of neck hackle makes the main colour appear wholly orange-red, the black parts being scarcely visible. Back red, shading to orange towards saddle hackle. Wing bows mahogany-red, each feather tipped with white; wing bars russet-red with lustrous green-black pea-shaped spots at ends, finishing with silvery white triangular tips, the whole forming regular bars across the wings.

Primaries black with a thin edging of chamois on outside; the visible lower third of each secondary feather chamois, upper two-thirds black. Remainder of wing a uniform chamois, each feather having at end a large pea-shaped white spot on a black triangle, the tips spaced evenly to conform with shape and outline of wing. (Note the reversal of these pattern-markings from the normal arrangement.) Tail feathers black with a metallic-green lustre, having a fine edging or lacing of dark chamois, and terminating with a white triangle. Breast, foot feathering and remainder of plumage throughout of golden-chamois ground colour, each feather having a light chamois shaft and finished with a black pea-shaped spot tipped with a white triangle.

Female plumage: Ground colour uniform golden-chamois, each feather terminating with a black pea-shaped spot tipped with a white triangle. Tail feathers black, finely laced with chamois and with white tips. Wings markings and other plumage as described for male, allowing for natural sexual differences.

In both sexes: Eyes orange-red with black pupils. Beak and nails slate-blue. Legs and feet slate-blue.

Defects to be avoided: Ground colour too light or washed out. White markings excessively gay or unevenly distributed.

The porcelaine

Male and female plumage: This is an extraordinarily delicate colour pattern. Markings and patterns generally are as described for millefleur in both sexes, with the exception that ground colour is light straw and the pea-shaped spots are pale blue, tipped with white triangles. Pale blue is substituted for the black of the millefleur in both sexes.

In both sexes: Eyes orange-red with black pupils. Beak and nails slate-blue. Legs and feet slate-blue.

Defects to be avoided: Ground colour too light or washed out. White markings too gay or unevenly distributed. Bareness or very poor quality feathering across the wing bows of the male.

The quail

Male plumage: This is a very striking colour scheme, with head feathers dark green-black, finely laced with gold. Crest of Watermael black ground colour with buff lacing and buff shafts. Beard golden-buff or nankin, shading darker towards the eyes, where plumage is black, finely laced with gold. Neck hackle with brilliant black ground, sharply laced with buff, having a golden lustre and yellowish-buff shafts. Back black ground colour with gold lacing, starting in middle of feathers and narrowing towards the tips, forming lance-like points with golden silky barbs and well defined light ochre-coloured shafts from root to point. These feathers are relatively broad under the neck, but narrower and longer nearing the saddle hackle. Colour more intense and black ground more pronounced towards the saddle hackle. Wing bows light gold, lower half of each feather black and clearly defined from the upper half, which should be nankin; wing bars light ochre, each feather having black triangular tip, the triangles forming two regular bars across the wing; bottom third of secondaries chamois colour, other two-thirds dull black; primaries dull black, hidden when the wing is closed. Tail black with metallic-green lustre, finely bordered with brown and with faintly defined light shafts. Sickles black, side hangers black, laced with chamois, with well-defined light shafts. Breast nankin, each feather finely laced with ochre (yellowish-buff), the shafts being distinct and clear. Thighs same colour as breast, abdomen and underparts greyish-brown, with silky, golden barb-shaped tips.

The general effect is that in this variety all the upper parts are dark and the lower parts light, giving the appearance of being covered with a dark chequered cloak. The dominating dark tint is chocolate-black, with a soft silvery lustre, known amongst artists as 'umber'. The general light tone is nankin or yellow-ochre, and well defined light shafts are important.

Female plumage: Head, face and neck covered with feathers which increase in size as they near the body, ground colour umber with very fine gold lacing. Crest of Watermael umber ground colour, with gold lacing and shafts. Neck velvety, darker than the back and clearly detached from it. Shaft and lacing clearer and more golden towards the breast. Back covered with umber-coloured feathers having a silvery velvety lustre, each feather dark, finely laced with chamois and with bright nankin shafts showing in strong contrast. Wings the same colour as back, dark umber finely laced with chamois, feathers broader and brighter towards lower part of wing; primaries are hidden when wing is closed and are dark, intense umber. Tail plumage and cushion similar to back and of same character. Breast, clear even nankin, the shafts pale and distinct, feathers nearing the wings finely and progressively bordered with dark umber, forming a distinctive colour-pattern.

In both sexes: Eyes dark brown (nearly black) with black pupils. Legs and feet slate grey. Beak and nails horn.

Defects to be avoided: Salmon or brownish breast.

The blue quail

Male and female plumage: Similar to the quail in all respects except that black markings are replaced by blue.

The silver quail

Male plumage: Head covered with feathers of a dark greenish-black, finely laced with white. Beard, white, going darker towards the eyes where the feathers assume a black ground colour finely laced with white. Neck hackle, silky feathers with a brilliant black ground sharply laced with white, and having a light-coloured shaft. Breast, deep solid white with shafts being very distinct and clear. Back, black to umber with a lacing of white, which starts at the middle of the feather and gets narrower towards the top, forming a lance-like point with white barbs, clearer than the lacing, ending in the upper part of the feather. The shaft well defined. Light colour divides the feather from the root to the point. These feathers are relatively broader under the neck hackle, becoming narrower and longer. Towards the saddle hackle, the colours become more intense, the black ground more noticeable in proportion as it approaches the end of the saddle hackle feathers. Wings, bows white, lower half of each feather black to umber and laced with white; wing bars, white each feather having a black triangular tip, the triangles forming two regular bars across the wing; bottom third of secondaries white, other two-thirds dull black; primaries dull black hidden when wing is closed. Tail, black with metallic-green lustre finely bordered with dull black/umber and with finely defined light shafts. Sickles black. Thighs, same as breast – white. Abdomen and underparts, greyish-white with silky white barb-shaped tips.

The general effect is the same as for the normal, blue and lavender quail. The upper parts are dark and the lower parts are light, giving the appearance of being covered with a dark chequered cloak. The dominating dark tint is chocolate/black with a soft silvery lustre known as 'umber'. The light tone is white and well-defined light shafts are important.

Female plumage: Head, face and neck ground colour umber with fine white lacing. Beard as in male. Neck velvety, darker than the back and clearly detached from it. Shaft and lacing clearer and more white towards the breast. Breast clear even solid white. The shafts pale and distinct. Feathers nearing the wings finely and progressively bordered with dark umber, forming a distinctive pattern. Back covered with umber-coloured feathers having a silvery lustre, each feather dark, finely laced with white, with white/bright shafts showing in stronger contrast. Wings same as the back, dark umber finely laced with white feathers broader and brighter towards the lower part of the wing; primaries are hidden when the wing is closed and are dark intense umber. Tail and cushion, similar to back and of same character.

In both sexes: Eyes dark brown (nearly black) with black pupils. Legs and feet slate-grey. Beak and nails horn.

Defects to be avoided: Salmon, brown or nankin on breast.

The lavender quail

In this colour the dark upper parts of the normal quail are replaced by pale silvery blue (uniform throughout the body) and the lower light parts are replaced by straw varying to cream according to the area of the body and the sex.

Male plumage: Head feathers laced with cream. Beard cream, darkening to straw towards the eyes, laced with gold. Neck hackle sharply laced in cream with golden lustre with lightish shafts. Back cream laced, golden-straw barbs, cream shafts. Wing bows light cream. The lower part of the feather clearly defined from the upper half which should be straw. Wing bars light ochre; bottom third of secondaries, cream. Tail bordered with golden-straw. Side hangers laced with chamois. Breast cream laced with ochre with distinct shafts. Thighs same as the breast. Abdomen and underparts silver/blueish grey with straw barb-shaped tips.

Female plumage: Head, face and neck cream laced. Beard as in the male. Breast cream, shafts pale and distinct. Back feathers laced with cream/straw with light shafts. Wings same as back laced with cream, going lighter towards bottom of the wing. Tail plumage and cushion similar to back.

In both sexes: Eyes orange-red with black pupils. Beak, nails, legs and feet slate-blue.

Defects to be avoided: Salmon, brown, nankin or white on breast.

The cuckoo

Male and female plumage: Uniformly cuckoo coloured, with transverse bands of dark bluish-grey on light grey ground. Each feather must have at least three bands.

In both sexes: Eyes orange-red. Legs, feet, beak and nails white, often spotted with bluish-grey in young birds.

Defects to be avoided: Feathers white or spotted with white, excessive number of black feathers, red on shoulders, wings and hackle.

The black mottled

Male and female plumage: All feathers black with metallic-green lustre, regularly tipped with white, tips varying in size with the feather. Excessive white markings or uneven distribution to be avoided.

In both sexes: Eyes dark red. Legs and feet slate-blue or blackish. Beak and nails dark horn.

The black

Male and female plumage: Black all over with metallic-green lustre, avoiding false colouring.

In both sexes: Eyes black, in Watermael dark brown. Legs and feet blue (blackish in young birds). Beak and nails black or very dark horn.

The white

Male and female plumage: Clear white throughout, avoiding false colours, straw tinge or yellow tint on back.

In both sexes: Eyes orange-red. Legs, feet, beak and nails white.

The lavender, or Renold's blue

Male and female plumage: This is a true-breeding pale silvery blue, all the feathers being of one uniform shade.

In both sexes: Eyes orange-red with black pupils. Beak and nails slate-blue. Legs and feet slate-blue.

Defects to be avoided: Any straw colouring in the hackles of the male.

The blue

Male plumage: Hackles, saddle, wing bow, back and tail dark slate-blue, also crest of Watermael. Remainder medium slate-blue, each feather to show lacing of darker shade.

Female plumage: Medium slate-blue with darker lacing all through, except head and neck, dark slate-blue.

In both sexes: Beak dark slate or horn. Eyes dark red or red-brown. Legs and feet slate. Nails dark slate or horn.

Other colours

These include blue mottled (similarly marked to black mottled); ermine (black-pointed white); fawn ermine (black-pointed fawn or pale buff); partridge; silver and gold.

Not all these colours are seen regularly in this country, but there is a practically no limit to the sub-varieties capable of being produced in these very charming breeds.

BARBU D'EVERBERG (RUMPLESS D'UCCLE)

BARBU DU GRUBBE (RUMPLESS D'ANVERS)

The above two types should follow the types in every respect except for the following.

General characteristics: male

Tail completely absent, the whole of the lower back being covered with saddle feathers.

Female

The general characteristics are similar to those of the male, allowing for the natural sexual differences.

Colour

As for all Belgian Bantam types.

Disqualifications

Any sign of tail.

BOOTED BANTAM

Origin: Europe
Classification: True bantam: Rare
Egg colour: Tinted

The Booted Bantam, or Sabelpoot as it is called in the Netherlands is an ancient breed. In the early part of this century it was crossed with the Barbu d'Anvers (Antwerp Bearded Bantam) to create the Belgian Barbu d'Uccle bantam which it resembles. The main difference between the breeds is that there is no muffling whatever in the Booted, showing large round wattles and having a narrower neck characteristic.

General characteristics: male

Carriage: Erect and strutting.
Type: Body short and compact. Full and prominent breast. Short back, the male's furnished with long and abundant saddle feathers. Large, long wings, carried in a drooping fashion. Large tail, full and upright, the male's sickles a little longer than the main feathers and slightly curved, coverts long, abundant, and nicely curved.
Head: Skull small. Beak rather stout, of medium length. Eyes bright and prominent. Comb single, small, firm, perfectly straight and upright, well serrated. Face of fine texture, free from hairs. Ear-lobes small and flat. Wattles small, fine, and well rounded.
Neck: Rather short, with full hackle.
Legs and feet: Short. Thighs well feathered at the hocks. Fairly short shanks heavily furnished on the outer sides with long and rather stiff feathers, those growing from the hocks almost touching the ground. Toes, four, straight and well spread, the outer and the middle being very heavily feathered.
Plumage: Long and abundant.

Female

The general characteristics are similar to those of the male, allowing for the natural sexual differences.

Colour

The black

Male and female plumage: Black, as lustrous as possible.
In both sexes: Beak black or horn. Eyes dark red or very dark brown. Comb, face, wattles and ear-lobes bright red. Legs and feet black.

The white

Male and female plumage: Pure snow-white.
In both sexes: Beak white. Eyes red. Comb, face, wattles and ear-lobes bright-red. Legs and feet white.

The black mottled, millefleur and porcelaine: Colour as for Belgian bantams.

Weights

Male 850 g (30 oz)
Female 750 g (27 oz)

White Booted Bantam male

Scale of points

Colour (of plumage 20, legs and beak 10)	30
Leg and foot feathering	15
Type	15
Head	15
Size	15
Condition	10
	100

Serious defects

Other than single comb, or four toes on each foot. Any deformity.

BRAHMA

LARGE FOWL

Origin: Asia
Classification: Heavy: Soft feather
Egg colour: Tinted

Although the name Brahma is taken from the river Brahmaputra in India, it is now generally agreed that they were created in America from large feather legged birds imported from China in the 1840s known as Shanghais. These were crossed with Malay type birds from India, known as Grey Chittagongs, which introduced the pea comb and the beetle brow.

Rivalry between breeders of various strains led to a wide variety of names and much

Dark Brahma female, large

Buff Columbian Brahma female, large

Gold Brahma male, large

confusion. A panel of judges meeting in Boston, USA, in 1852 declared the official name to be Brahmapootras later shortened to Brahma. After a consignment of nine birds was sent to Queen Victoria in 1852, the Brahma became one of the leading Asiatic breeds in this country. Both light and pencilled Brahmas were included in the Poultry Club's first Book of Standards in 1865.

General characteristics: male

Carriage: Sedate, but fairly active.
Type: Body broad, square and deep. Back short, either flat or slightly hollow between the shoulders, the saddle rising halfway between the hackle and the tail until it reaches the tail coverts. Breast full, with horizontal keel. Wings medium sized with lower line horizontal, free from twisted or slipped feathers, well tucked under the saddle feathers, which should be of ample length. Tail of medium length, rising from the line of the saddle and carried nearly upright, the quill feathers well spread, the coverts broad and abundant, well curved, and almost covering the quill feathers.
Head: Small, rather short, of medium breadth, and with slight prominence over the eyes. Beak short and strong. Eyes large, prominent. Comb triple or 'pea', small, closely fitting and drooping behind. Face smooth, free from feathers or hairs. Ear-lobes long and fine, free from feathers. Wattles small, fine and rounded, free from feathers.
Neck: Long, covered with hackle feathers that reach well down to the shoulders, a depression being apparent at the back between the head feathers and the upper hackle.
Legs and feet: Legs moderately long, powerful, well apart, and feathered. Thighs large and covered in front by the lower breast feathers. Fluff soft, abundant, covering the hind parts, and standing out behind the thighs. Hocks amply covered with soft rounded feathers, or with quill feathers provided they are accompanied with proportionately heavy shank and foot feathering. Shank feather profuse, standing well out from legs and toes, extending under the hock feathers and to the extremity of the middle and outer toes,

profuse leg and foot feather without vulture hock being desirable. Toes, four, straight and spreading.
Plumage: Profuse, but hard and close compared with the Cochin.

Female

With the exception of the neck and legs, which are rather short, the general characteristics are similar to those of the male, allowing for the natural sexual differences.

Colour

The dark

Male plumage: Head silver-white. Neck and saddle hackles silver-white, with a sharp stripe of brilliant black in the centre of each feather tapering to a point near its extremity and free from white shaft. Breast, underpart of body, thighs and fluff intense glossy black. Back silver-white, except between the shoulders where the feathers are glossy black laced with white. Wing bows silver-white; primaries black, mixed with occasional feathers having a narrow white outside edge; secondaries, part of outer web (forming 'bay') white, remainder ('butt') black; coverts glossy black, forming a distinct bar across the wing when folded. Tail black, or coverts laced (edged) with white. Leg feathers black, or slightly mixed with white.
Female plumage: Head silver-white or striped with black or grey. Neck hackle similar to that of the male, or pencilled centres. Tail black, or edged with grey, or pencilled. Remainder any shade of clear grey finely pencilled with black or a darker shade of grey than the ground colour, following the outline of each feather, sharply defined, uniform, and numerous.

The light

Male plumage: Head and neck hackle as in the dark variety. Saddle white preferably, but white slightly striped with black in birds having very dark neck hackles. Wing primaries black or edged with white; secondaries white outside and black on part of inside web. Tail black, or edged with white. Remainder clear white, with white, blue-white, or slate undercolour, not visible when the feathers are undisturbed. Black-and-white admissible in toe feathering. Shank feathers white.
Female plumage: Neck hackle silver-white striped with black (dense at the lower part of the hackle), the black centre of each feather entirely surrounded by a white margin. In other respects the colour of the female is similar to that of the male.

The white

Male and female plumage: Pure white throughout.

The gold

Male plumage: Head rich gold. Neck and saddle hackles, rich gold, with sharp stripe of brilliant black in the centre of each feather tapering to a point near its extremity and free from gold shaft. Breast, underparts of body, thighs and fluff intense glossy black. Back rich gold except between the shoulders where the feathers may be laced with gold. Wing bows bright red; primaries black, with a narrow outer edge of rich bay; secondaries, outer web (forming 'bay') partly bay, free from outer edge of black, remainder (forming 'butt') black. Wing coverts glossy black, forming a distinct bar across the wing when folded. Tail black or coverts edged with gold. Footings and leg feathers black, or slightly mixed with gold.
Female plumage: Head rich gold or striped with black. Neck hackle rich with sharp brilliant black striping free from shaftiness, the striping completely surrounded by gold. Tail black, edged with gold or pencilled. Remainder of plumage rich, even, clear gold, finely pencilled with black; the markings numerous, sharply defined and uniform, following the outline of the feather.

The buff columbian

Male plumage: Head golden-buff. Neck and saddle hackles golden-buff, with a sharp

stripe of brilliant black in the centre of each feather tapering to a point near its extremity and free from buff shaft. Saddle golden-buff preferably, but slightly striped with black in birds having very dark neck hackles. Wing primaries black or edged with golden-buff; secondaries golden-buff outside and black on part of inside web. Tail black, or edged with golden-buff. Remainder a clear golden-buff, with buff or slate undercolour. Black and buff in toe feathering. Shank feather buff.
Female plumage: Neck hackle golden-buff striped with black (dense at the lower part of the hackle), the black centre of each feather entirely surrounded by a golden-buff margin. In other respects the colour of the female is similar to that of the male.

In both sexes and all colours: Beak yellow or yellow and black. Eyes orange-red. Comb, face, ear-lobes and wattles bright red. Legs and feet orange-yellow or yellow.

Weights

Male 4.55–5.45 kg (10–12 lb)
Female 3.20–4.10 kg (7–9 lb)

Scale of points

Type, size and carriage	35
Colour (including purity and brilliance), markings and feather	30
Legs and feet (including foot feather and leg colour)	15
Head and eye	15
Condition	5
	100

Serious defects

Comb other than 'pea' type. Badly twisted hackle or wing feathers. Total absence of leg feather. Great want of size in adults. Total want of condition. White legs. Any deformity. Buff on any part of the plumage of light. Much red or yellow in the plumage, or much white in the tail of dark males. Utter want of pencilling, or patches of brown or red in the plumage of dark females. Split or slipped wings.

BANTAM

Brahma bantams are exact miniatures of their large fowl counterparts so all standard points apply.

Weights

Male 1080 g (38 oz) max.
Female 910 g (32 oz) max.

BRAKEL

LARGE FOWL

Origin: Belgium
Classification: Light: Rare
Egg colour: White

What has been said of the Campine is more or less true of the Brakel, which, from the evidence obtainable, was descended many centuries ago from the same stock. The name was given for the reason that they have been bred extensively in Belgium in the neighbourhood of Nederbraekel in Flanders. The main difference from the Campine (which is

Silver Brakel male, bantam

Gold Brakel male, large

hen feathered in the male) is that the male Brakel has full flowing saddle hackles (of the ground colour) and large well-curved sickles, also a larger, deeper body with much coarser and broader barring. Once bred in a number of colours, but now the leading varieties are the gold and silver. Brakel were first imported into Britain in 1898.

General characteristics: male

Carriage: Alert and graceful.
Type: Body proportionately long and deep, back moderately long, wide at the shoulders, narrowing to base of the tail, sloping slightly from shoulders to tail. Breast full, wide, deep, well rounded, carried well forward but not beyond a line drawn perpendicular with tip of beak. Wings moderately long and large, carried high, well above lower thighs, tips concealed by saddle feathers and ending short of stern. Main tail feathers broad, long, carried well back at an angle of 45° and well spread, sickles broad, long and well curved, lesser sickles, coverts and saddle hackles long and abundant.
Head: Medium size, broad and deep. Beak medium length slightly curved. Comb single, upright, of medium size evenly serrated, the back carried well out just clear of the neck, slightly coarse in texture. Eyes full, round and bright. Face smooth and free from wrinkles. Ear-lobes medium sized, oval shaped and smooth. Wattles medium size, well rounded, free from wrinkles or folds.
Neck: Moderately long, gracefully arched and well covered with hackle feathers.
Legs and feet: Moderately long shanks, round and free from feathers. Toes, four, medium length and well spread.

Female

With the exception of the single comb which falls gracefully over one side of the face, the general characteristics are similar to those of the male allowing for the natural sexual differences.

The gold

Male plumage: Head, back and saddle hackles rich golden-bay, remainder beetle-green black barring on rich golden-bay ground. Each feather must be barred in a transverse direction with the end of the feather golden-bay, the bars being clear and with well-defined edges, running across each feather so as to form, as near as possible, rings around the body. The markings on the tail, wings, thighs and fluff are black, being twice as wide as the ground colour. The black on the breast and body is of equal width with the ground colour.
Female plumage: With the exception of the back and saddle which should be barred with the black being twice the width of the ground colour, the rest of the plumage is as the male.

The silver

Male plumage: Head, neck hackle, back and saddle hackles pure white, remainder beetle-green black barring on pure white ground, the markings being as in the gold.
Female plumage: Head and neck hackle pure white, remainder beetle-green black barring on pure white ground, the markings being as in the gold.

In both sexes and colours

Beak horn. Eyes dark brown with black pupil. Comb, face and wattles bright-red. Ear-lobes white. Legs and feet leaden blue. Toenails horn. Undercolour slate.

Weights

Male 3.20 kg (7 lb)
Female 2.75 kg (6 lb)

Scale of points

Type	20
Markings	20
Colour	20
Head	15
Size	10
Legs and feet	10
Condition	5
	100

Serious defects

Lack of barring on breast, indistinct barring on female's back. White in face. Lack of size in adult birds. Excess ticking in neck hackle. Male's comb twisted, oversize or falling over. Dark pigmentation in comb of female. Any deformity.

BANTAM

Brakel bantams should follow exactly the standard for the large fowl.

Weights

Male 790 g (28 oz)
Female 700 g (24 oz)

BREDA

LARGE FOWL

Origin: Netherlands
Classification: Light: Rare
Egg colour: White or cream

This breed, also having the names Guelderlanders and Kraaikoppen (Dutch for crow-headed) has been known at least as far back as the 1840s. Although not widely bred outside their homeland Bredas have been described in many poultry books as they have a unique feature in having no comb at all.

General characteristics: male

Carriage: Moderately upright.
Type: The body and wings of a typical medium-weight breed, back fairly long. Breast rather full and well rounded. Tail large, full and well fanned, with the top line of the sickles carried at about 45° above the horizontal at the junction with the back.
Head: Beak rather long, strong, well hooked. Nostrils with a long structure similar to that found on crested breeds. Comb entirely absent. The feathers on top of the head usually stand upright, but there must be no crest. Ear-lobes and wattles moderately large.
Neck: Medium length, well rounded and furnished.
Legs and feet: Legs of above average length, with long vulture hocks. Feet have four toes with moderate feathering down shanks and on middle and outer toes.
Plumage: Moderately abundant, but not too fluffy.

Female

Similar to the male, allowing for natural sexual differences. However carriage is lower, and the breast is more prominent and rounded.

Black Breda male

Colours

In all varieties, face and lobes bright red. Ear-lobes white. Eyes reddish-brown to orange-brown.

The black

Male and female plumage: A clear glossy green-black. Beak horn colour/black mixed. Feet black or slate-grey.

The laced blue

Male and female plumage: A clear bluish-grey with dark slate lacing and glossy dark slate or black hackles. Beak horn. Feet slate.

The blue

Male and female plumage: A bright clear blue/grey as free as possible from lacing. Beak horn. Feet slate.

The cuckoo

Male and female plumage: Evenly banded dark grey on silvery white. Wholly white feathers to be severely penalized. Beak white to horn shaded. Feet white to light blue/grey.

The white

Male and female plumage: Pure white throughout. Beak white to horn shaded. Feet white to light blue-grey.

Weights

Cock	3 kg ($6\frac{1}{2}$ lb) or over;	cockerel	2.5 kg ($5\frac{1}{2}$ lb) or over
Hen	2.25 kg (5 lb) or over;	pullet	1.75 kg (4 lb) or over

Scale of points

Head points	30
Type, carriage	20
Legs and feet	20
Colour	15
Condition	15
	100

Serious defects

Any deformity. Any comb or crest development.

BANTAM

The description is as for large fowl.

Weights

Cock 1000 g (36 oz); cockerel 900 g (32 oz)
Hen 800 g (29 oz); pullet 700 g (24 oz)

CAMPINE

LARGE FOWL

Origin: Belgium
Classification: Light: Rare
Egg colour: White

The Campine (pronounced Kam-peen), having the same ancestry as the Brakel, has been known in Belgium since the reign of Emperor Charles V. Its name comes from the area in which it was bred, the dry sandy plains around Antwerp. The stock which came to England in the nineteenth century did not have the same full tails as the Brakel and so were improved along those lines. Thus, 'hen feathered' males were evolved. Body type is different to the Brakel, and having Asiatic blood has made for a deeper bodied bird with full, flowing hackles. The gold and silver are the only accepted colours for the Campine whereas the Brakel has had more colours standardized. A breed club was formed in 1899 which drew up a standard where exhibition males and females could be produced from one breeding pen. The Campine has played a major role in the work of autosex-linkage. The curiosity of the genetic workers at Cambridge was aroused by the fact that the Campine is not, as might be supposed, a barred breed like the barred Plymouth Rock. The difference led to the making of the autosexing Cambar.

General characteristics: male

Carriage: Alert and graceful.
Type: Body broad, close and compact. Back rather long, narrowing to the tail. Breast full and round. Wings large and neatly tucked. Tail carried fairly high and well spread. Campine males are hen feathered, without sickles or pointed neck and saddle hackles. The top two tail feathers slightly curved.
Head: Moderately long, deep, and inclined to width. Beak rather short. Eyes prominent. Comb single, upright, of medium size, evenly serrated, the back carried well out and clear of neck; free from excrescences. Face smooth. Ear-lobes inclined to almond shape, medium size, free from wrinkles. Wattles long and fine.
Neck: Moderately long and well covered with hackle feathers. The formation of the neck feathers in the Campine is called the cape.

Campine large fowl
Silver male
Silver female

Legs and feet: Legs moderately long. Shanks and feet free from feathers. Toes, four, slender and well spread.

Female

With the exception of the single comb which falls gracefully over one side of the face, the general characteristics are similar to those of the male, allowing for the natural sexual differences.

Colour

The gold

Male and female plumage: Head and neck hackle rich gold, not a washed-out yellow. Remainder beetle-green barring on rich gold ground. Every feather must be barred in a transverse direction with the end gold, the bars being clear and with well-defined edges, running across the feather so as to form, as near as possible, rings round the body and three times as wide as the ground (gold) colour. On the breast and underparts of the body the barrings should be straight or slightly curved; on the back, shoulders, saddle and tail they may be of a V-shaped pattern, but preferably straight.

The silver

Male and female plumage: Head and neck hackle pure white. Remainder beetle-green barring on pure white ground, the markings being as in the gold.

In both sexes and colours

Beak horn. Eyes dark brown with black pupil. Comb, face and wattles bright-red. Ear-lobes white. Legs and feet leaden blue. Toenails horn.

Weights

Male	2.70 kg (6 lb)
Female	2.25 kg (5 lb)

Scale of points

Size	10
Head (comb 5, eyes 5, lobes 5)	15
Colour (cape 12, sheen 10)	22
Tail (development and carriage)	8
Legs and feet	5
Markings	30
Condition	10
	100

Interpretation

The ideal is a bird clearly, distinctly and evenly barred all over with the sole exception of its neck hackle, which should be of the ground colour of the body. So that, taking the five main points of the bird – viz. neck hackle, top (including back, shoulders and saddle), tail, wing and breast – each is of as much importance as another, and judges are requested to bear in mind that a specimen excelling in one or two particulars but defective in others should stand no chance against one of fair average merit throughout. Special attention should be paid to size, type and fullness of front in breeding and judging Campines.

Serious defects

Sickle feathers or pointed hackles on the males. Bars and ground colour of equal width. Ground colour pencilled. Comb at the back too near the neck. Side spikes (or sprigs) on comb. Legs other than leaden-blue. White in face. Red eyes. Feather or fluff on shanks. Any deformity. Dark pigmentation in combs of females. White toenails. Slate-blue beak. Black around the eyes.

BANTAM

Campine bantams should follow exactly the standard for the large fowl.

Weights

Male 680 g (24 oz)
Female 570 g (20 oz)

COCHIN

LARGE FOWL

Origin: Asia
Classification: Heavy: Soft feather
Egg colour: Tinted

The Cochin, as we know it today, originally came from China in the early 1850s, where it was known as the Shanghai, and later still as the Cochin-China. The breed created a sensation in this country in poultry circles because of its immense size and table properties. Moreover, it was an excellent layer. It was developed, however, for wealth of feather and fluff for exhibition purposes to the extent that its utility characteristics were neglected, if not made impossible, in winning types. There are no Cochin bantams.

General characteristics: male

General shape and carriage: Massive and deep. Carriage rather forward, high at stern, and dignified.
Type: Body large and deep. Back broad and very short. Saddle very broad and large with a gradual and decided rise towards the tail forming an harmonious line with it. Breast broad and full, as low down as possible. Wings small and closely clipped up, the flights being neatly and entirely tucked under the secondaries. Tail small, soft, with as little hard quill as possible, and carried low or nearly flat.
Head: Small. Beak rather short, curved and very stout at base. Eyes large and fairly prominent. Comb single, upright, small perfectly straight, of fine texture, neatly arched and evenly serrated, free from excrescences. Face smooth, as free as possible from feathers or hairs. Ear-lobes sufficiently developed to hang nearly or quite as low as the wattles, which are long, thin and pendant.
Neck: Rather short, carried somewhat forward, handsomely curved, thickly furnished with hackle feathers which flow gracefully over shoulders.
Legs and feet: Thighs large and thickly covered with fluffy feathers standing out in globular form; hocks entirely covered with soft curling feathers, but as free as possible from any stiff quills (vulture hocks). Shanks short, stout in bone, plumage long, beginning just below hocks and covering front and outer sides of shanks, from which it should be outstanding, the upper part growing out from under thigh plumage and continuing into foot feathering. There should be no marked break in the outlines between the plumage of these sections. Toes, four, well spread, straight, middle and outer toes heavily feathered to ends. Slight feathering of other two toes is a good sign to breeders.

Female

With certain exceptions the general characteristics are similar to those of the male, allowing for the natural sexual differences. Comb and wattles as small as possible. The body more square than the male's and the shoulders more prominent. The back very flat, wide and short, with the cushion exceedingly broad, full and convex, rising from as far forward as possible and almost burying the tail. Wings nearly buried in abundant body

Partridge Cochin female

Blue Cochin female

Buff Cochin female

feathering and the tail very small. Breast full, as low as possible. General shape is 'lumpy', massive and square. Carriage is forward, high at cushion, with a matronly appearance.

Colour

The black

Male and female plumage: Rich black, well glossed, free from golden or reddish feathers.

In both sexes: Beak yellow, horn or black. Comb, face, ear-lobes and wattles bright-red. Eyes bright red, dark red, hazel or nearly black. Legs dusky yellow or lizard.

The blue

Male plumage: Hackle, back and tail level shade of rich dark blue free from rust, sandiness or bronze. Remainder even shade of blue free from lacing on breast, thighs or fluff and free from rust, sandiness or bronze.

Female plumage: One even shade of blue free from lacing; pigeon blue preferred.

In both sexes: Beak yellow, horn or yellow slightly marked with horn. Comb, face, ear-lobes and wattles bright red. Eyes dark. Legs and feet blue with yellow tinge in pads.

The buff

Male plumage: Breast and underparts any shade of lemon-buff, silver-buff or cinnamon provided it is even and free from mottling. Head, hackle, back, shoulders, wings, tail and saddle may be any shade of deeper and richer colour which harmonizes well – lemon, gold, orange or cinnamon – wings to be perfectly sound in colour and free from mealiness. White in tail very objectionable.

Female plumage: Body all over any even shade, free from mottled appearance. Hackle of a deeper colour to harmonize, free from black pencilling or cloudiness, cloudy hackles being especially objectionable. Tail free from black.

In both sexes: Beak rich yellow. Comb, face, ear-lobes and wattles brilliant red. Eyes to match plumage as nearly as possible, but red eyes preferred although rare. Legs bright yellow with shade of red between the scales.

The cuckoo

Male and female plumage: Dark blue-grey bands across the feather on blue-grey ground, the male's hackle free from golden or red tinge, and his tail free from black or white feathers.

In both sexes: Beak rich bright yellow, but horn permissible. Comb, face, ear-lobes and wattles as in the black. Eyes bright red. Legs brilliant yellow.

The partridge and grouse

Male plumage: Neck and saddle hackle rich bright red or orange-red, each feather with a dense black stripe. Back, shoulder coverts and wing bow rich red, of a more decided and darker shade than the neck. Wing coverts green-black, forming a wide and sharply cut bar across the wing; secondaries rich bay outside the black inside, the end of every feather black; primaries very dark bay outside and dark inside. Saddle rich red or orange-red, the same colour as, or one shade lighter than, the neck. Remainder glossy black, as intense as possible, white in tail objectionable.

Female plumage: Neck bright gold, rich gold, or orange-gold, with a broad black stripe in each feather, the marking extending well over the crown of the head. Remainder (including leg feathering) brown (darker in the grouse), distinctly pencilled in crescent form with rich dark brown or black, the pencilling being perfect and solid up to the throat.

In both sexes: Beak yellow or horn. Comb, face, ear-lobes and wattles as in the black. Eyes bright red. Legs yellow, but may be of a dusky shade.

The white

Male and female plumage: Pure white, free from any straw or red shade.

In both sexes: Beak rich bright yellow. Comb, face, ear-lobes and wattles as in the black. Eyes pearl or bright red. Legs brilliant yellow.

Weights

Cock	4.55–5.90 kg (10–13 lb);	cockerel	3.60–5.00 kg (8–11 lb)
Hen	4.10–5.00 kg (9–11 lb);	pullet	3.20–4.10 kg (7–9 lb)

Scale of points

Feathering (cushion 8, fluff 7, tail 5, hackle 5, legs 10)	35
Colour (or markings in cuckoo or partridge)	20
Size	15
Head (ear-lobes 5)	15
Type	10
Condition	5
	100

Serious defects

Primary wing feathers twisted on their axes. Utter absence of leg feather. Badly twisted or falling comb. Legs other than yellow or dusky yellow, except in blacks and blues. Black spots in buffs. Brown mottling (if conspicuous) in partridge males, or pale breasts destitute of pencilling in partridge females. White or black feathers in cuckoos. Crooked back, wry tail or any other deformity.

CRÈVECOEUR

LARGE FOWL

Origin: France
Classification: Heavy: Rare
Egg colour: White

The Crèvecoeur is an old French breed not unlike an Houdan, but having a horn-type comb, four toes and a heavier, broad, square-built body with bold upright carriage.

General characteristics: male

Carriage: Bold and upright.
Type: Body large, broad and practically square. Well-rounded breast. Flat back. Large well-folded wings. Full tail carried moderately high.
Head: Skull large, with a decidedly pronounced protuberance on top. Crest full and compact, round on top and not divided or split, inclined slightly backwards fully to expose the comb, not in any way obstructing the sight except from behind, composed of feathers similar to those of the hackles, and the ends almost touching the neck. Beak strong and well curved. Eyes full. Comb of the horn type (V-shaped), of moderate size, upright and against the crest, each branch smooth (free from tines) and tapering to a point. Face muffled, the muffling full and deep, extending to the back of the eyes, hiding the lobes and the face. Ear-lobes small but not exposed. Wattles moderately long.
Neck: Long and graceful, thickly furnished with hackle feathers.
Legs and feet: Wide apart, short and the shanks free from feathers. Toes, four, straight and long.

Female

With the exception of the crest, which must be of globular shape and almost concealing the comb, the general characteristics are similar to those of the male, allowing for the natural sexual differences.

Colour

Male and female plumage: Lustrous green-black. No other colour is admissible, except a few white feathers in the crests of adults, which, however are not desirable.

In both sexes: Beak dark horn. Eyes bright red, although black is admissible. Comb, face, wattles and ear-lobes bright red. Legs and feet black or slate-blue.

Weights

Male	4.10 kg (9 lb)
Female	3.20 kg (7 lb)

Scale of points

Crest and muffling	30
Size	20
Comb	15
Colour	15
Type	10
Condition	10
	100

Serious defects

Coloured feathers. Loose crest obstructing the sight. Any deformity.

BANTAM

Crèvecoeur bantams should follow the large fowl standard.

Weights

Male	1020 g (36 oz)
Female	790 g (28 oz)

CROAD LANGSHAN

LARGE FOWL

Origin: Asia
Classification: Heavy: Soft feather
Egg colour: Brown

The first importation of Langshans into this country was made by Major Croad and, as with other Asiatic breeds, controversy centred around it. Already there was the black Cochin and then the black Langshan, some contending both were one breed, and others that they were quite separate Chinese breeds. As developed here the breed was called the Croad Langshan after the name of the importer. In 1904, a Croad Langshan club was formed to maintain the original stamp of bird. The Modern Langshan has been developed along different lines and, in consequence, the two types are shown in separate classes at shows.

General characteristics: male

Carriage: Graceful, well balanced, active and intelligent.
Type: Back of medium length, broad and flat across shoulders, the saddle well filling the angle between the back and the tail as seen in profile. In the male the back should appear shorter than in the female. Breast broad, deep and full (fuller in old bird) with long breastbone, the keel slightly rounded. Wings carried high, and well tucked up. Tail fan-shaped, well spread to right and left and carried rather high; it should be level with the head when the bird stands in position of attention; side hangers plentiful, and two sickle feathers on each side projecting some 15 cm (6 in) or more beyond rest. Abdomen capacious and resilient to touch, with fine pelvic bones. Saddle rather abundantly furnished with hackles.
Head: Carried well back, small for size of bird, full over the eyes. Beak fairly long and slightly curved. Eyes large and intelligent. Comb single, upright, straight, medium or rather small, free from side sprigs, thick and firm at the base, becoming rather thin, fine and smooth in texture, evenly serrated with five or six spikes (five preferred). Face free of feathers. Ear-lobes well developed, pendant, and fine in texture. Wattles fine in quality and rather small.
Neck: Of medium length, with full neck hackle.
Legs and feet: Legs sufficiently long to give a graceful carriage to the body which should be well balanced, an adult bird neither high nor low on leg. Thighs rather short but long enough to let the hocks stand clear of fluff, well covered with soft feathers. Shanks medium length, well apart, feathered down outer sides (neither too scantily nor too heavily). Toes, four, long, straight and slender, the outer toe feathered.
Plumage: Rather soft, neither loose nor tight.
Table merits: Size for table purposes must be a great consideration, consistent with type. Bone medium or rather fine, in due proportion to size, but subordinate to amount of meat carried. Male has higher proportion of bone than female. Skin, thin and white. Flesh white.

White Croad Langshan male, large

Black Croad Langshan female, large

Female

Hocks need not show in adult as she carries more fluff than the male. Cushion fairly full but not obtrusive. Tail may have two feathers slightly curved and projecting about 2.5 cm (1 in) beyond rest. In other respects similar to the general characteristics of the male, allowing for the natural sexual differences.

Colour

Male and female plumage: Surface dense black with beetle-green gloss free from purple or blue tinge. Undercolour dark grey, darker in the female. White in foot feather characteristic and not a defect.

In both sexes: Beak light to dark horn, preferably light at tip and streaked with grey. Eyes brown, the darker the better, but not black (ideal is colour of ripe hazel nut, a Vandyke brown). Comb, face, wattles and ear-lobes brilliant red. Shanks bluish black (bluish in adult birds, scales and toes nearly black in young birds) showing pink between scales especially on back and inner side of shank. In male bird intense red should show through the skin along outer side at base of shank feathers. Toes, the web and bottom of foot pinkish white, the deeper the pink the better; black spots on soles a serious fault. Toenails white, dark colour or black a serious fault.

White Langshans (Croad type)

The general characteristics are the same as for the original Croad above. Plumage is pure white, and the beak light horn. Eyes, comb, wattles and legs are as in the original Croad. Serious defects are black or coloured feathers, and black tips to the feathers.

Weights

Male 4.10 kg (9 lb) min.
Female 3.20 kg (7 lb) min.

Scale of points

Type and condition (shape 15, condition 10)	25
Body (girth 15, frame and bone 10)	25
Plumage (colour 15, furnishings and footings 10)	25
Head, feet and abdomen (head and feet 15, abdomen and pelvis 10)	25
	100

Note: It is left to discretion of judge to penalize any bad fault to the extent of 25 points.

Disqualifications

Yellow legs or feet. Yellow in face at base of beak or in edge of eyelids. Five toes. Other than single comb. Permanent white in ear-lobes. Grey (light slate colour) in webbing of flights. Black or partly black soles of feet as distinct from black spots. Vulture hocks.

Not objectionable (in stock birds and not seriously against birds in show-pen, judges using their discretion)

Purple or blue barring in few feathers only where others are of good colour. Dark red in few feathers in neck hackle or on shoulders of male. White (not grey) in flights and secondaries. White tips on head of adult female. White tips or edging on breast in chicken feathers. Moderate amount of feathering on middle toe.

Highly objectionable (to be firmly discouraged)

Appreciable amount of purple or blue barring. Decided purple or blue tinge. Light eye (make some allowance for age). Yellow iris. Wry or squirrel tail. Marked scarcity or absence of leg and foot feather.

BANTAM

Croad Langshan bantams to follow large fowl standard.

Weights

Male 770–910 g (27–32 oz)
Female 650–790 g (23–28 oz)

DOMINIQUE

LARGE FOWL

Origin: America
Classification: Heavy: Rare
Egg colour: Brown

This is perhaps the oldest of the distinctive American breeds, being mentioned in the earliest poultry books as an indigenous and valued variety, as an excellent layer, very hardy and good for the table. They were first seen in this country at the Birmingham Show of 1870 and reimported in 1984.

General characteristics: male

Carriage: Erect and graceful, denoting an active fowl.
Type: Body broad, full, compact. Back medium length, moderately broad rising with concave sweep to the tail. Breast broad, round and carried well up. Wings rather large, well-folded and carried without drooping, the ends being covered by the saddle hackle. Tail long, full, slightly expanded, carried at an angle of 45° above the horizontal, main tail broad and overlapping, sickles long, well-curved.
Head: Medium size, carried well up. Beak short, stout, well-curved. Face surface smooth, skin fine and soft in texture. Eyes large, full and prominent. Comb rose, not so large as to overhang the eyes or beak, firm and straight on the head, square in front, uniform on both sides, free from hollow centre, terminating in a spike at the rear, the point of which turns slightly upward, the top covered with small points. Ear-lobes oval, of medium size. Wattles broad, medium in length, well rounded, smooth, fine in texture, free from folds or wrinkles.
Neck: Medium length, well-arched, tapering. Hackle abundant.
Legs and feet: Medium length legs, set well apart, straight when viewed from in front, shanks fine, round and free from feathers. Toes four, well spread.
Plumage: Feathering soft but close, body fluff moderately full.

Female

The general characteristics are similar to those of the male, allowing for the natural sexual differences.

Colour

Male and female plumage: Slate. Feathers in all sections of the bird crossed throughout the entire plumage by irregular dark and light bars that stop short of positive black and white, the tip of each feather dark, free from shafting, brownish tinge or metallic sheen, excellence to be determined by distinct contrasts. The male may be one or two shades lighter than the female.

In both sexes: Beak, legs and feet yellow. Eyes reddish-brown. Face, comb, ear-lobes and wattles bright red. Undercolour slate.

Dominique male

Weights

Male	2.72–3.17 kg (6–7 lb)
Female	1.81–2.26 kg (4–5 lb)

Scale of points

Type	20
Colour	20
Markings	15
Head	15
Legs and feet	10
Tail	7
Size	7
Condition	6
	100

Serious defects

Coarse, over refined and/or crow head. Low wing carriage. Flat and cutaway breast. Shallow and narrow bodied. Red or yellow in any part of plumage, completely white feathers. Any physical deformity.

Dominique female

BANTAM

Dominique bantams should follow exactly the standard for large fowl.

Weights

Male	793 g (28 oz)
Female	566 g (20 oz)

DORKING

LARGE FOWL

Origin: Great Britain
Classification: Heavy: Soft feather
Egg colour: Tinted

Its purely British ancestry makes the Dorking one of the oldest of domesticated fowls in lineage. A Roman writer, who died in AD 47, described birds of Dorking type with five

Silver grey Dorking male, large

Silver grey Dorking female, large

Red Dorking female, large

toes, and no doubt such birds were found in England by the Romans under Julius Caesar. By judicious crossings, and by careful selection, the Darking or Dorking breed was established.

General characteristics: male

Carriage: Quiet and stately, with breast well forward.
Type: Body massive, long and deep, rectangular in shape when viewed sideways, and tightly feathered. Back broad and moderately long with full saddle inclined downward to the tail. Breast deep and well rounded with a long straight keel bone. Wings large and well tucked up. Tail full and sweeping carried well out (a squirrel tail being objectionable) with abundant side hangers and broad well-curved sickles.
Head: Large and broad. Beak stout, well proportioned and slightly curved. Eyes full. Comb single or rose. Either kind is allowed in darks, single only in reds and silver greys, and rose only in cuckoos and whites. The single comb is upright, moderately large, broad at base, evenly serrated, free from thumb marks or side spikes. The rose is moderately broad and square fronted, narrowing behind to a distinct and slightly upturned leader, the top covered with small coral-like points of even height, free from hollows. Face smooth. Ear-lobes moderately developed and hanging about one-third the depth of the wattles, which are large and long.
Neck: Rather short, covered with abundant hackle feathers falling well over the back, making it appear extremely broad at the base, and tapering rapidly at the head.
Legs and feet: Legs short and strong. Thighs large and well developed but almost hidden by the body feathering. Shanks short, moderately stout and round (square or sinewy bone being very objectionable), free from feathers, the spurs set on the inner side and pointing inwards. Toes, five, large, round and hard ('spongy' feet to be guarded against), the front toes (three) long, straight and well spread, the hind toe double and the

extra toe well formed, viz. the normal toe as nearly as possible in the natural position, and the extra one placed above, starting from close to the other, but perfectly distinct and pointing upwards.

Female

The general characteristics are similar to those of the male, allowing for the natural sexual differences, except that the tail is carried rather closely. The single comb, too, falls over one side of the face.

Colour

The cuckoo

Male and female plumage: Dark grey or blue bands on light blue-grey ground, the markings uniform, the colours shading into each other so that no distinct line or separation of the colours is perceptible.

The dark

Male plumage: Hackles (neck and saddle) white or straw more or less striped with black. Back various shades of white, black and white or grey, mixed with maroon or red (bronze objectionable). Wing bows white, or white mixed with black or grey; coverts (or bar) black glossed with green; secondaries outer web white, inner black. Breast and underparts jet black; white mottling not permissible. Tail richly glossed black, and a little white on primary sickles is permissible, but white hangers decidedly objectionable.

Female plumage: Neck hackle white or pale straw, striped with black or grey-black. Breast salmon-red, each feather tipped with dark grey verging on black. Tail nearly black, the outer feathers slightly pencilled. Remainder of plumage nearly black, or approaching a rich dark brown, the shaft showing a cream-white, each feather slightly pale on the edges, except on the wings, where the centre of the feather is brown-grey covered with a small rich marking surrounded by a thick lacing of the black, and free from red. Another successful colour is every feather over the body pencilled a brown-grey in the centre, with lacing round, and the breast as described above.

The red

Male plumage: Hackles (neck and saddle) bright glossy red. Back and wing bows dark red. Remainder of plumage jet black glossed with green.

Female plumage: Hackle bright gold heavily striped with black. Tail and primaries black or very dark brown. Remainder of plumage red-brown, the redder the better, each feather more or less tipped or spangled with black, and having a bright yellow or orange shaft.

The silver grey

Male plumage: Hackles (neck and saddle) silver-white free from straw tinge or marking of any kind. Back, shoulder coverts and wing bow silver-white free from striping. Wing coverts lustrous black with green or blue gloss; primaries black with a white edge on outer web; secondaries white on outer and black on inner web, with a black spot at the end of each feather, the corner of the wing when closed appearing as a bar of white with a black upper edge. Remainder of plumage deep black, free from white mottling or grizzling, although in old males a slight grizzling of the thighs is not objectionable.

Female plumage: Hackle silver-white, striped with black. Breast robin red or salmon-red ranging to almost fawn, shading off to ash-grey on the thighs. Body clear silver-grey, finely pencilled with darker grey (the pencilling following the outer line of the feather), free from red or brown tinge or black dapplings.

Note: The effect may vary from soft dull grey to bright silver-grey, an old fashioned grey slate best describing the colour. Tail darker grey, inside feathers black.

The white

Male and female plumage: Snow-white, free from straw tinge.

In both sexes and all colours

Beak white or horn, dark horn permissible in the dark. Eyes bright red. Comb, face, wattles and ear-lobes brilliant red. Legs and feet (including nails) a delicate white with a pink shade.

Weights

Cock 4.55–6.35 kg (10–14 lb); cockerel 3.60–5.00 kg (8–11 lb)
Hen 3.60–4.55 kg (8–10 lb)

Scale of points

	Dark	*Silver grey or red*	*Cuckoo or white*
Size	28	18	15
Type	20	12	20
Colour	12	24	15
Fifth toe	10	10	15
Condition	12	12	10
Head	10	16	17
Feet, condition of	8	8	8
	100	100	100

Serious defects

Total absence of fifth toe. Legs other than white or pink-white, or with any sign of feathers. Spurs outside the shank. Single comb in cuckoo or white. Rose comb in red or silver-grey. White in breast or tail of silver-grey male. Any coloured feathers in white. Very long legs. Crooked or much swollen toes. Bumble feet. Any deformity.

BANTAM

Standards for large fowl to be used for Dorking bantams.

Weights

Male 1130–1360 g (40–48 oz)
Female 910–1130 g (32–40 oz)

DUTCH BANTAM

Origin: The Netherlands
Classification: True bantam
Egg colour: Tinted

The Dutch Bantam (or De Hollandse Krielan) in its country of origin has been around for a long time, though in Holland a club was only formed on 1 December 1946. The breed first appeared in this country around the late 1960s, and a club was formed in 1982. Since then the breed has gone from strength to strength, with thirteen colours standardized, though in Holland many more varieties keep appearing.

General characteristics: male

Carriage: Upright and jaunty.
Type: Back very short, broad at shoulders, slightly sloping, saddle short and broad with abundant hackle running smoothly into tail coverts. Breast carried high, full and well

Blue partridge Dutch male

Yellow partridge Dutch female

Gold partridge Dutch female

forward. Wings relatively large and long, but not too pointed, carried low and close. Tail upright, full and well spread, with well-developed and curved sickles.
Head: Small, face smooth. Comb single, small with five serrations tending towards flyaway type. Beak short and strong, slightly curved. Eyes large and lively. Wattles fine, short and round. Ear-lobes small and fine, oval to almond shape.
Neck: Short, curved and finely tapered with plentiful hackle.
Legs and feet: Legs well spaced and straight, thighs short, shanks short and free of feather. Toes four, well spread.
Plumage: Luxuriant and lying close to body with plentiful sickles, side hangers and coverts.

Female

The general characteristics are similar to those of the male, allowing for the natural sexual differences.

Colour

The gold partridge (black-red)

Male plumage: Head orange reddish-brown. Neck hackle a gradual transition from orange to light orange-yellow, each feather having a greenish-black middle stripe. Back deep reddish-brown. Side hackle corresponding with neck hackle, a little darker permitted. Breast black with green sheen, free from markings or spots. Wing bow black, shoulders deep reddish-brown, wing bar iridescent greenish-black; primaries inner web and tip black, outer web chestnut-brown; secondaries inner web black, outer web chestnut-brown; wing bay chestnut-brown when wing is closed. Thighs deep black with green sheen, free from markings or spots. Abdomen black. Tail main feathers, sickles and tail coverts green iridescent black, the tail coverts nearest the side hangers with a brownish edge underneath at the tip. Undercolour greyish.

Female plumage: Head gold-brown. Throat greyish-white. Neck hackle goldish-yellow, with a black middle stripe. Wing, back, saddle and tail coverts greyish-brown with fine black peppering, as even as possible, free of rust or red. Tail feathers blackish, the top feather on each side with brown peppering. Breast light salmon-brown, shading to brownish-grey near the thighs. Thighs and down ash-grey.

In both sexes. Beak dark horn or bluish. Legs and feet slate-blue.

Faults: Any mismarked feathers. Any splashing or coloured feathers in black parts of male. Rusty colour in wings of female.

The silver partridge (silver duckwing)

Male plumage: Exactly the same as the gold partridge male in the black feathered parts and in the markings on the neck and saddle hackles. The orange and light orange-yellow and the brown replaced by silvery white.

Female plumage: Head silver-white. Hackle silver-white, each feather having a black middle stripe. Wing coverts, back, saddle and talk coverts muted silver or slate-grey with fine black peppering as even as possible, free from flecks, rust-brown or yellow. Tail feathers blackish, the top feather on each side with silver peppering. Breast light salmon-brown, fading to ash-grey underneath. Rump and rear underparts ash-grey. Thighs ash-grey with some peppering.

In both sexes: Beak dark horn or bluish. Legs and feet slate-blue.

Faults: Any mismarked feathers. Any splashing or coloured feathers in black parts of male. Red in plumage of male. Rusty colour in wings of female.

The yellow partridge (yellow duckwing)

Male plumage: Head rich straw-yellow. Neck hackle light straw-yellow, each feather having a greenish-black middle stripe. Back deep golden-orange. Saddle hackle corresponding with neck hackle, a little darker permitted. Breast black with green sheen free from markings and spots. Wing bow black. Shoulders deep golden-orange. Wing bar iridescent greenish-black; primaries inner web and tip black, outer web creamy yellow-straw; secondaries inner web black, outer web creamy yellow-straw; wing bay being creamy yellow-straw when wing is closed. Thighs deep black with green sheen, free from markings and spots. Abdomen black. Tail main sickles and tail coverts green iridescent black, the tail coverts nearest the side hangers with a straw-coloured edge underneath at tip. Undercolour greyish.

Female plumage: Head greyish-yellow. Throat yellowy white. Neck hackle straw-yellow with a black middle stripe. Wing, back, saddle and tail coverts greyish straw-yellow with fine black peppering, as even as possible, free from red rust. Tail main feathers blackish, the top feather on each side with yellow peppering. Breast rich salmon-yellow fading to greyish yellow underneath. Rump and rear underparts grey-yellow. Thighs grey-yellow with some peppering.

In both sexes: Beak dark horn or bluish. Legs and feet slate-blue.

Faults: Any mismarked feathers. Any splashing or coloured feathers in black parts of male. Red or silver in plumage of male. Any rust or silver in wings of female.

The blue silver partridge (blue silver duckwing)

The blue yellow partridge (blue yellow duckwing)

Male plumage: In both colours, exactly the same as the silver partridge and yellow partridge with the exception that all black parts are replace with a clear even blue colour free from lacing.

Female plumage: In both colours, exactly the same as the silver partridge and yellow partridge with the exception that all black is replaced with a clear even blue, free from lacing, this includes the peppering.

In both sexes and colours: Beak dark horn or bluish. Legs slate-blue.

Faults: Any mismarked feathers, red in plumage of male blue silver partridge, or red or silver in plumage of male blue yellow partridge. Rusty coloured wings in blue silver partridge females or rust or silver in wings of blue yellow partridge female. Wrong coloured legs.

The blue partridge (blue-red)

Male plumage: Exactly the same as the partridge male with the exception that all black parts are replaced with a clear even blue colour, free from lacing.

Female plumage: Exactly the same as the partridge female with the exception that all black is replaced with a clear, even blue; this includes peppering.

In both sexes: Beak dark horn or bluish. Legs and feet slate-blue.

Faults: Any mismarked feathers. Any splashing or coloured feathers in the blue parts of the male. Rusty colour in wings of the female.

The red shouldered white (pyle)

Male plumage: Neck hackle orange-red, each feather having a white centre stripe. Back and shoulders carmine-red. Saddle hackle orange-red, each feather having a white centre stripe. Wing bar white-cream; primaries white-cream; secondaries outer web chestnut, inner web white-cream, the chestnut only showing when wing closed. Remainder of plumage white-cream.

Female plumage: Neck hackle gold-yellow, each feather having a white centre stripe. Breast salmon. Remainder of plumage white-cream.

In both sexes: Beak bluish-white or horn. Legs and feet light slate-blue.

Faults: Any splashing or coloured feathers in white parts. Coloured parts of male lacking in depth of colour. Breast colour of female too pale and body colour other than white-cream.

The cuckoo partridge (crele)

Male plumage: Neck hackle straw colour banded with gold or black. Back and shoulders bright gold-chestnut banded with straw-yellow. Wing bar dark grey banded with pale grey; primaries and secondaries dark grey banded with pale grey; outer web of secondaries chestnut, the chestnut only showing when wing closed. Saddle hackle pale straw banded gold. Breast and underparts dark grey banded with light grey. Tail and tail coverts dark grey banded with light grey.

Female plumage: Neck hackle pale gold banded with black. Breast salmon. Body greyish-brown with indistinct soft banding. Wings dark greyish-brown with slightly lighter banding. Tail dark grey-black with slightly lighter bands.

In both sexes: Beak bluish-horn. Legs light slate-blue or pearl-grey.

Faults: White feathers in tail or wings.

The cuckoo

Male and female: Light blue-grey ground colour with each feather banded across with broad bands of dark blue-grey. Banding to be distinct and uniform. In the male a lighter shade is permissible provided the banding is distinct. Undercolour banded but of a lighter shade. Beak light horn or bluish. Legs and feet white.

Faults: Black or white feathers. Unclear or uneven banding.

The black

Male and female plumage: Rich glossy black with beetle-green sheen throughout. Undercolour dark. Beak dark horn or bluish. Legs and feet dark slate.

Faults: Any white feathers. Purple banding. Light undercolour. Red in hackle of males. Tendency to gypsy face. Wrong leg colour.

The white

Male and female plumage: Pure snow-white throughout, free from any yellow or straw tinge or any black splashes. Beak white or bluish. Legs slate-blue or light slate-blue.

Faults: Any yellow or straw tinge or any black splashes or peppering. Serious fault is white legs.

Salmon Faverolles male, large

Salmon Faverolles female, large

Cuckoo Faverolles female, bantam

wing bows bright cherry-mahogany. In older males a fringe of dark orange on the wing bows is permissible. Breast, thighs, underfluff, tail and shank feathering black. Wing bar black; primaries black; secondaries white outer edge, black inner edge and at tips. Straw colouring on wing bows is objectionable and serious defects are extremely pale shoulders and wing bows of almost silver duckwing colouration and very dark heavily striped neck hackles of almost partridge colouration.

Female plumage: Beard and muffs creamy white. Breast, thighs and underfluff cream. Remainder wheaten-brown, head and neck striped with dark shade of the same colour but free from black. Wings similar to back but softer and lighter; primaries, secondaries and tail wheaten-brown. (Typical undercolour on the top of the body is grey in the upper part of the fluff, fading towards the root. On the front and underside it is mostly creamy white. The usual colour of the inner edge of the primaries, secondaries and tail feathers is grey.) A dark, blotchy breast is objectionable, as is white in wing feathers and extremely pale back colour with almost white wing bows.

In both sexes: Beak, etc. as in the buff.

The white

Male and female plumage: Pure white throughout.

In both sexes: Beak, etc. as in the buff.

Weights

Cock	4.00–5.00 kg (9–11 lb);	cockerel	3.40–4.50 kg ($7\frac{1}{2}$–10 lb)
Hen	3.40–4.30 kg ($7\frac{1}{2}$–$9\frac{1}{2}$ lb);	pullet	3.20–4.00 kg (7–9 lb)

Scale of points

Utility/table qualities, size and condition	25
Type	25
Colour	20
Beard and muffling	15
Formation of feet and toes	5
Foot feather (shanks and outer toes)	5
Comb	5
	100

Serious defects

Skin and legs other than white (except in the black, blue and cuckoo). Narrowness or in-kneed. Hollow breast. White or brassiness in hackle, wing or saddle, or purple barring, or white in foot feather in the blue. Mealiness of general colour, or white in tail, wings and undercolour of the buff. Other than white legs, smuttiness on back, in the ermine. Brassiness of wings of white male. Yellow or red in hackle of male cuckoo.

Disqualifications

Absence of muffling. Crooked breastbone or any bodily deformity. Other than five toes on each foot. Featherless shanks and outer toes.

BANTAM

Faverolles bantams to be exact replicas of their large fowl counterparts (the scale of points differs slightly).

Weights

Male 1130–1360 g ($2\frac{1}{2}$–3 lb)
Female 900–1130 g (2–$2\frac{1}{2}$ lb)

Scale of points

Utility/table qualities, size and condition	20
Type	25
Colour	20
Beard and muffs	20
Formation of feet and toes	5
Foot feather (shanks and outer toes)	5
Comb	5
	100

It is accepted that a specimen excelling in one or two main areas but defective in other important features should stand no chance against a bird of fair average merit throughout.

FAYOUMI

LARGE FOWL

Origin: Egypt
Classification: Light: Rare
Egg colour: White or cream

The Fayoumi is an ancient Egyptian breed from the district of Fayoum and has been

Fayoumi female

selectively bred for egg production. They are hardy, very early maturing, strong fliers and vocal when handled. The Fayoumi is not genetically a barred breed but a pencilled breed. All chicks are born brown whether of the silver or gold variety. The plumage pattern is similar to that of the Brakel. They were introduced into the UK in 1984.

General characteristics: male

Carriage: Alert, graceful, very active, rather upright.
Type: Body wedge-shaped, of moderate length and good depth, back moderately long, rather broad at shoulders narrowing towards tail. Breast moderately full and nicely rounded. Wings of moderately length carried well above lower thigh. Saddle full and flowing. Tail abundant, fairly large, well spread and carried high, but not squirrel.
Head: Medium size, rather broad. Beak medium length, strong and slightly curved. Eyes large, full and bright. Comb single, of medium size and evenly serrated. Face smooth and fine, free from wrinkles and folds. Ear-lobes elongated oval, smooth and fine. Wattles of medium length, well rounded.
Neck: Moderately long, gracefully arched. Hackle abundant, flowing down over back and shoulders.
Legs and feet: Thighs short and strong. Shanks medium length, round, free from feathers. Toes four, of medium length and well spread.

Female

The female has a smaller, upright comb and allowing for natural sexual differences, the general characteristics are similar to the male.

Colour

The silver pencilled

Male plumage: Head, hackles, back and saddle, pure silver-white. Saddle coverts black

laced with silver. Tail solid black with beetle-green sheen. Remainder of plumage pure silver-white with coarse imprecise beetle-green black barring. Thighs and fluff barred. Every feather ends with a silver tip. Finer pencilling on wings to form a bar.
Female plumage: Head and hackle pure silver-white. Tail darkly barred or pencilled, main tail black. Remainder of plumage silver-white ground colour with coarse, imprecise beetle-green black barring. The bars appearing to form irregular rings around the body, approximately three times as wide as the ground colour. Thighs and fluff are barred. Every feather ending in a silver tip. Finer pencilling on wings to form a bar.

The gold pencilled

Male and female plumage: Except that the ground colour is gold, this variety is similar to the silver.

In both sexes and colours

Beak horn. Eyes dark brown. Comb, face wattle and ear-lobes bright-red. Legs and feet slate-blue. Toenails horn.

Weights

Cock	1.81 kg (4 lb);	cockerel	1.36 kg (3 lb)
Hen	1.36–1.58 kg (3–3½ lb);	pullet	900 g–1130 g (2–2½ lb)

Scale of points

Type	30
Head	10
Colour	10
Legs	10
Condition	10
Tail	10
Size	10
Carriage	10
	100

Serious defects

White ear-lobes. Black striping in neck. Laced body feathers. Any deformities in comb. Any specimen that does not bear the features of a productive layer with alert carriage.

BANTAM

Fayoumi bantams should follow exactly the standard for large fowl.

Weights

Male	430 g (16 oz)
Female	400 g (14 oz)

FRIESIAN

LARGE FOWL

Origin: Holland, province of Friesland
Classification: Light: Rare
Egg colour: White

One of the oldest Dutch breeds and recognized as rare in Holland. Used in the past as a good laying breed, this quality is still maintained.

General characteristics: male

Carriage: Upright, bold and active.
Type: Body broad at the shoulders and moderately long, back sloping down slightly and narrowing to the tail. Breast carried high, full and forward. Back appears shorter with the high tail carriage. Wings are long and substantial and carried close above the thigh, slightly sloping backwards. Tail is large and well spread and carried high. Saddle feathers well developed.
Head: Small to medium sized, somewhat oblong, fine hairy feathers on cheeks allowed. Comb single, small to medium in size, erect and evenly serrated (five or six), following the line of the skull but standing well clear of the neck. Eyes large. Beak medium length, slightly curved. Ear-lobes oval, small. Wattles medium length, round.
Neck: Long, slender near the head.
Legs and feet: Medium length legs, set well apart. Four well spread toes.
Plumage: Full but close fitting.

Female

The abdomen is very well developed. The end of the comb may droop to one side. The other characteristics are the same as for the male, allowing for the natural sexual differences.

Colour

The gold pencilled

Male plumage: Head rich bright bay. Neck hackle, back and saddle rich lustrous bay. Front of neck rich reddish-bay. Main tail black. Sickles and lesser sickles lustrous greenish-black. Coverts greenish-black, edged with reddish-bay. Wing shoulders, front and bows lustrous reddish-bay; wing coverts rich bay, forming a distinct bar across the wing; primaries black, lower edge of lower feathers edged with bay; secondaries outer webs bay with black marking, inner web black with bay marking, exposed portion of webs forming wing bay, bay. Breast rich reddish-bay. Body light bay. Abdomen fluff bay, shading to light bay at rear, some black marking. Lower thighs reddish-bay. Undercolour slaty blue.
Female plumage: Head bright reddish-bay. Neck hackle bright bay. Front of neck golden-buff. Back, cushion, tail coverts and wings golden-buff marked with parallel rows of elongated black ovals, each oval extending slightly diagonally across web but not to edge. (The elongated ovals might also be called elongated spangles. The pencilling favours the description of the Sicilian Buttercup and is really not barred but has two or three pairs of small black spangles arranged in a parallel fashion.) Main tail black, lower webs barred with buff marking. Primaries buff splashed with black; secondaries golden-buff, barred with parallel black marking on outer web, forming wing bay, inner webs black edged with golden-buff. Upper breast golden-buff sparsely marked with small black ovals. Lower breast, body and lower thighs golden-buff marked same as back. Abdomen fluff buff. Undercolour slaty blue.

The silver pencilled

The same as the gold pencilled except that the reds, bays and buffs are replaced with white.

The chamois pencilled

The same as the gold pencilled except that all the dark parts are a rich yellow/buff and all the other parts are white.

In both sexes and all colours

Face, comb, wattles red. Eyes dark orange. Beak horn. Ear-lobes pure white. Legs slate-blue.

Weights

Male 1.40–1.60 kg (3–3½ lb)
Female 1.20–1.40 kg (2¾–3 lb)

Scale of points

Type	20
Colour and markings	30
Head	15
Condition	20
Legs and feet	5
Tail	10
	100

Serious defects

Too small, narrow frame, squirrel tail, tail carried too low. Fully folded comb in female. Black/brown eye colour.

BANTAM

Bantam Friesian are an exact miniature of the large fowl counterpart.

Weights

Male 550–650 g (19–22 oz)
Female 450–550 g (16–19 oz)

FRIZZLE

LARGE FOWL

Origin: Asia
Classification: Heavy: Soft feather
Egg colour: White or tinted

The Frizzle, a purely exhibition breed, is of Asiatic origin, and is notable for its quaint feather formation, each feather curling towards the head of the bird. It is more popular in bantams than in large fowls.

General characteristics: male

Carriage: Strutting and erect.
Type: Body broad and short. Breast full and rounded. Wings long. Tail rather large, erect, full but loose, with full sickles and plenty of side hangers. Lyre tails in males desirable but not obligatory.
Head: Fine. Beak short and strong. Eyes full and bright. Comb single, medium sized and upright. Face smooth. Ear-lobes and wattles moderate size.
Neck: Of medium length, abundantly frizzled.
Legs and feet: Legs of medium length. Shanks free from feathers. Toes, four, rather thin, and well spread.
Plumage: Moderately long, broad and crisp, each feather curled towards the bird's head, and the frizzling as close and abundant as possible.

Female

The general characteristics are similar to those of the male, allowing for the natural sexual differences, except that the comb is much smaller and the neck is not so abundantly frizzled.

Colour

Male and female plumage: Black, blue, buff or white, a pure even shade throughout in the 'self coloured' varieties; columbian as in Wyandotte; duckwing, black-red, brown-red, cuckoo, pyle and spangle as in Old English Game; red as in Rhode Island Red.

In both sexes and all colours

Beak yellow in the buff, columbian, pyle, red and white varieties; white in the spangle, black-red and cuckoo; and dark willow, black or blue in other varieties. Eyes red. Comb, face, wattles and ear-lobes bright red. Legs and feet to correspond with the beak. (There are variations in leg colour, and yellow legs are frequently demanded in blacks, though not so standardized.)

Weights

Cock	3.60 kg (8 lb);	cockerel	3.20 kg (7 lb)
Hen	2.70 kg (6 lb);	pullet	2.25 kg (5 lb)

Scale of points

Type	25
Colour	25
Curl of feather	30
Condition	10
Weight	10
	100

Serious defects

Narrow feather. Want of curl. Long tail. Drooping comb. Other than single comb. White lobes. Deformity of any kind.

BANTAM

Frizzle bantams follow the large fowl standard although a slightly different scale of points is used.

Weights

Male	680–790 g (24–28 oz)
Female	570–680 g (20–24 oz)

Scale of points

Head and comb	5
Legs and feet	5
Plumage colour	15
Size	10
Curl	25
Feather quality	20
Type and symmetry	10
Condition	10
	100

Blue Frizzle female, bantam

White Frizzle female, bantam

Cuckoo Frizzle male, bantam

GERMAN LANGSHAN

LARGE FOWL

Origin: Germany
Classification: Heavy: Rare
Egg colour: Brown to brown-yellow

The German Langshan originally came from northern China. The first birds were imported to Germany from England in 1879, when the breed quickly grew. The breeding of the smooth-legged form succeeded exclusively in Germany. In other countries, clean-legged Langshans were either imported from Germany or arose accidently. The imported Langshan formed the basis for German breeding. Sporadically, other breeds were crossed in, particularly the Black Minorca, but the pure blood has always predominated. The new form, which differs from the old in that the back is no longer carried horizontally, has been bred since 1904 by decree of the German Club of Langshan Breeders.

General characteristics: male

Carriage: An upstanding, elegant and beautiful bird, with long extended body, leaning slightly forwards and with rising back-line.
Type: Body strong, full and equally broad from front to back. Breast broad, arched and fairly deep. Belly broad, deep, full and downy. Back long, lowest bearing starting directly under the neck hackle, from there, broad but not angular shoulders, the same width up the tail with smooth lines. Wings firmly set but carried high. Tail short, but carried in a continuing line with the back, the coverts broad, soft and abundant, curved and covering the tail.

Head: Small, fairly narrow and slightly arched. Beak moderately long, strong and slightly curved. Eyes large and prominent. Comb small, single, erect, slightly arched and evenly serrated. Face smooth and slightly hairy. Wattles small, oval or round, of fine texture. Ear-lobes fairly long, narrow and slender.
Neck: Long, slightly curved, with moderately long hackles.
Legs and feet: Legs long, not too thick, shanks unfeathered. Toes four, long, thin, straight and well spread with strong claws.

Female

The female displays the form of the male in a graceful manner but without marked cushion formation and with roomy hindquarters. The somewhat more loosely carried tail stands out only slightly from the coverts. Otherwise the general characteristics are similar to those of the male, allowing for the natural sexual differences.

Colour

The black
Male and female plumage: Black with beetle-green sheen.

The white
Male and female plumage: Pure snow-white throughout.

The blue
Male and female plumage: As near to the colour of the Andalusian as possible.

In both sexes and all colours
Face, comb, wattles and ear-lobes red. Beak black in the black and the blue varieties, white in the white and light-coloured varieties. Eyes brown-black in the black and the blue varieties, orange-red in the white and light-coloured varieties. Legs and feet, in the first year the feet and toes of the black and the blue varieties are black, later they are slate-blue. In the white and light-coloured varieties they are flesh coloured. Soles of the feet are light.

Weights

Male 3–4.5 kg ($6\frac{1}{2}$–10 lb)
Female 2.5–3.5 kg ($5\frac{1}{2}$–$7\frac{3}{4}$ lb)
Egg weight minimum 56 g (2 oz)

Scale of points

Type and carriage	35
Head	10
Thighs, legs and feet	20
Tail	15
Size and condition	10
Colour and plumage	10
	100

Defects

Long pointed beak, body slightly too high or low in bearing. Flat breast. White in wings in coloured varieties.

Serious defects

Any deformity or weakness. Narrow breast. Little body depth. Stumpy loose feathers. Red, yellow or grey eyes in the black or the blue varieties. Saw-like comb. White in ear-lobes. Twisted feathers. Sloping shoulders with sharply set tail, cushion shape, or hollow or roach back. Coverts higher than tail feathers. In the black variety, dull colour and purple barring. Very light beak. In the blue, sooty tone or uneven colour, undefined lacing, very light beak and too light feet. In the white, any other features which are no snow-white. Feathered shanks.

BANTAM

Bantam German Langshans should follow exactly the standard for the large fowl. They come in the same colours with the addition of cuckoo, buff, birchen and brassy-backs.

Weights

Male	1000 g (35 oz)
Female	900 g (31 oz)
Egg weight	35 g ($1\frac{1}{5}$ oz)

HAMBURGH

LARGE FOWL

Origin: North Europe
Classification: Light: Soft feather
Egg colour: White

The origin of the Hamburgh is wrapped in mystery. The spangled were bred in Yorkshire and Lancashire three hundred years ago as Pheasants and Mooneys, and there is a book reference to black Pheasants in the North of England in 1702. In its heyday, the Hamburgh was a grand layer and must have played its part in the making of other laying breeds. However, its breeders directed it down purely exhibition roads, until today it is in few hands.

General characteristics: male

Carriage: Alert, bold and graceful.
Type: Body moderately long, compact, fairly wide and flat at the shoulders. Breast well rounded. Wings large and neatly tucked. Tail long and sweeping, carried well up (but avoiding 'squirrel' carriage), the sickles broad and the secondaries plentiful.
Head: Fine. Beak short, well curved. Eyes bold and full. Comb rose, medium size, firmly set, square fronted, gradually tapering to a long, finely ended spike (or leader) in a straight line with the surface and without any downward tendency, the top level (free from hollows) and covered with small and smooth coral-like points of even height. Face smooth and free from stubby hairs. Ear-lobes smooth, round and flat (not concave or hollow), varying in size according to the variety. Wattles smooth, round and of fine texture.
Neck: Of medium length, covered with full and long feathers, which hang well over the shoulders.
Legs and feet: Legs of medium length. Thighs slender. Shanks fine and round, free of feathers. Toes, four, slender and well spread.

Female

The general characteristics are similar to those of the male, allowing for the natural sexual differences.

Colour

The black

Male and female plumage: Rich black, with a distinct green sheen from head to tail, and especially on sickle feathers and tail coverts. Any approach to bronze or purple tinge or barring to be avoided.

Black Hamburgh female, large

Gold pencilled Hamburgh male, bantam

Silver spangled Hamburgh male, bantam

Silver spangled Hamburgh female, bantam

In both sexes: Beak black or dark horn. Comb, face and wattles red. Ear-lobes white. Legs and feet black.

The gold pencilled

Male plumage: Bright red-bay or bright golden-chestnut, except the tail, which is black, the sickle feathers and coverts being laced all round with a narrow strip of gold.
Female plumage: Ground colour similar to the general colour of the male, and, except on the hackle (which should be clear of all marking, if possible), each feather distinctly and evenly pencilled straight across with fine parallel lines of a rich green-black, the pencilling and the intervening colour to be the same width, while the finer and the more numerous on each feather the better.

In both sexes: Beak dark horn. Comb, face and wattles red. Ear-lobes white. Legs and feet lead-blue.

The silver pencilled

Male and female plumage: Except that the ground colour, and in the male the tail lacings, are silver, this variety is similar to the gold pencilled.

The gold spangled

Male plumage: Ground colour rich bright bay or mahogany; striping, spangling, tipping and tail rich green-black. Hackles and back, each feather striped down the centre. Wing bow dagger-shaped tips at the end of each feather; bars (two), rows of large spangles, running parallel across each wing with a gentle curve, each bar distinct and separate; secondaries tipped with large round spangles, forming the 'steppings'. Breast and underparts, each feather tipped with a round spot or spangle, small near the throat, increasing in size towards the thighs, but never so large as to overlap.
Female plumage: Ground colour and spangling are similar to those of the male. Hackle, wing bars and 'steppings' as in the male. Tail coverts black, with a sharp lacing or edging of gold on each feather. Remainder, each feather tipped with a spangle, as round as possible, and never so large as to overlap, the spangling commencing high up the throat.

In both sexes: Beak, comb, face, wattles, ear-lobes, legs and feet as in the pencilled varieties.

The silver spangled

Male plumage: Ground colour pure silver; spangling and tipping rich green-black. Hackles, shoulders and back, each feather marked with small, dagger-like tips. Wing bow dagger-shaped tips, increasing in size until they merge into what is known as the third bar; bars (two) and secondaries, breast and underparts similarly marked to those of the gold spangled variety. Tail ending with bold half-moon-shaped spangles; sickles with large round spangles at the end of each feather; coverts similar, though spangles not so big.
Female plumage: Ground colour and spangling similar to those of the male. Hackle marked from the head with dagger-shaped tips, which gradually increase in width until they merge into the spangles at the bottom. Wing secondaries as in the male, bars similar to those of the gold spangled female. Tail, each feather with a half-moon-shaped spangle at the end; coverts reaching halfway up the true tail feathers to form a row across the tail (each side) of round spangles. Remainder marked as in the gold female.

In both sexes: Beak, comb, face, wattles, ear-lobes, legs and feet as in the pencilled varieties.

In all colours

Eyes red or dark, dark preferred.

Weights

Male	2.25 kg (5 lb) approx.
Female	1.80 kg (4 lb) approx.

Scale of points

The black	*Male*	*Female*
Type, style and condition	15	15
Head (comb, face and ear-lobes 15 each)	45	45
Colour (legs 5)	25	35
Tail	15	5
	100	100

The pencilled	*Male*	*Female*
Type, style and condition	10	10
Head (comb, ear-lobes and face)	25	20
Colour (including legs)	30	10
Markings (back and cushion 15, breast and thighs 15, tail 15, wings 10, neck hackle 5)	—	60
Tail markings	35	—
	100	100

The gold spangled	*Male and Female*
Type, style and condition	10
Head (comb 10, ear-lobes 5, face 5)	20
Colour (including legs)	10
Markings (back and saddle 15, breast and thighs 15, wings 15, neck hackle 10, tail 5)	60
	100

The silver spangled	*Male*	*Female*
Type, style and condition	10	10
Head (comb 10, ear-lobes 10, face 5)	25	—
Head (comb 10, ear-lobes 5, face 5)	—	20
Colour (including legs)	10	10
Markings (tail 15, neck hackle 10, back and saddle 10, breast and thighs 10, wings 10)	55	—
Markings (back and cushion 15, neck hackle 15, tail 10, breast and thighs 10, wings 10)	—	60
	100	100

Serious defects

White face. Single comb. Red ear-lobes. Squirrel or wry tail. Any other deformity.

BANTAM

Hamburgh bantams follow the large fowl standard except that black is not standardized as a bantam colour. The scale of points used is different.

Weights

Male 680–790 g (24–28 oz)
Female 620–740 g (22–26 oz)

These weights are to be treated as maximums although the pencilled varieties are usually considerably larger.

Scale of points

Markings	60
Head, comb, face and lobes	20
Colour	10
Type, style and condition	10
	100

In pencilled males the points allotted to markings are divided between tail and colour.

HOUDAN

LARGE FOWL

Origin: France
Classification: Heavy: Rare
Egg colour: White

Introduced into England in 1850, the Houdan is one of the oldest French breeds, taking its name from the town of Houdan, and has been developed for table qualities. Developed here it was once classified as a heavy breed, but today is included in the category of light, non-sitting breeds. It is one of the few breeds carrying a fifth toe, a semi-dominant feature when crossed with other breeds.

General characteristics: male

Carriage: Bold and active.
Type: Body broad, deep and lengthy, as in the Dorking. Tail full with the sickles long and well arched.
Head: Fairly large, with a decidedly pronounced protuberance on top, and crested. Crest full and compact, round on top and not divided or 'split', composed of feathers similar to those of the hackle, inclining slightly backwards fully to expose the comb, in no way obstructing the sight except from behind. Beak rather short and stout, well curved, and with wide nostrils. Eyes bold. Comb leaf type, somewhat resembling a butterfly placed at the base of the beak, fairly small, well defined, and each side level. Face muffled; muffling large, full, compact, fitting around to the back of the eyes and almost hiding the face. Ear-lobes small, entirely concealed by muffling. Wattles small and well rounded, almost concealed by beard.
Neck: Of medium length, with abundant hackle coming well down on the back.
Legs and feet: Legs short and stout, well apart, free of feathers. Toes, five, similar to those of the Dorking.

Female

The general characteristics are similar to those of the male, allowing for the natural sexual differences, with the exception of the crest, which is full, compact and globular, not in any way obstructing the sight except from behind, and with the comb visible. Tail fairly full.

Colour

Male and female plumage: Glossy green-black ground with pure white mottles, the mottling to be evenly distributed, except on the flights and secondaries, and in the male on the sickles and tail coverts, which are irregularly edged with white. *Note:* In young Houdans, black generally preponderates but what mottling there is should be even and clear. Mottling becomes gayer with age.

In both sexes: Beak horn. Eyes red. Comb, face and wattles bright red. Ear-lobes white or tinged with pink. Legs and feet white mottled with lead-blue or black.

Houdan male, bantam

Weights

Male 3.20–3.60 kg (7–8 lb)
Female 2.70–3.20 kg (6–7 lb)

Scale of points

	Male	*Female*
Type	12	10
Size	18	20
Comb	15	8
Legs and feet	10	10
Colour	15	15
Crest	12	15
Muffling	8	12
Condition	10	10
	100	100

Serious defects

Red or straw-coloured feathers. Loose crest obstructing the sight. Spur outside the shank. Feathers on shanks or toes. Other than five toes on each foot. Any deformity.

BANTAM

Houdan bantams should follow the large fowl standard.

Weights

Male 680–790 g (24–28 oz)
Female 620–740 g (22–26 oz)

INDIAN GAME

LARGE FOWL

Origin: Great Britain
Classification: Heavy: Hard feather
Egg colour: Tinted

To Cornwall must go the credit for giving us the Indian Game. Breeds used in the make-up were the red Asil, black-breasted red Old English Game, and the Malay. The breed has been developed for its abundant quantity of breast meat, in which respect no other breed can equal it. When large table birds were the most popular in this country Indian Game males were chosen as mates for females of such table breeds as the Sussex, Dorking and Orpington, to produce extra large crosses. The females chosen for mating belonged to breeds possessing white flesh and shanks. Jubilee Indian Game are similar to Indians, but the lacing is white; in Indians it is black. The two varieties are often interbred.

General characteristics: male

Carriage: Upright, commanding and courageous, the black sloping downwards towards the tail. A powerful and broad bird, very active, sprightly and vigorous.
Type: Body very thick and compact and very broad at the shoulders, the shoulder butts showing prominently, but the bird must not be hollow backed, the body tapering towards the tail. Back flat and broad at the shoulders, but the bird must not be flat sided. Elegance is required with substance. Breast wide, fairly deep and prominent, but well rounded, and rising to the vent. Wings short and carried closely to the body, well rounded at the points, closely tucked at ends and carried rather high in front. Tail medium length with short narrow secondary sickles and tail coverts, close and hard; carriage drooping.
Head: Of medium length and thick, not so keen as in English Game, nor as thick as in the Malay; somewhat beetle browed but not nearly as much as in the Malay. Skull broad. Beak well curved and stout where set on the head giving the bird a powerful appearance. Eyes full and bold. Comb (in undubbed birds) pea type, i.e. three longitudinal ridges, the centre one being double height of those at sides, small closely set on the head. Ear-lobes and wattles small, smooth and of fine texture.
Neck: Of medium length and slightly arched; hackle short, barely covering base of the neck.
Legs and feet: Legs very strong and thick. Thighs round and stout, but not as long as in the Malay. Shanks short and well scaled. The length of shank must be sufficient to give the bird a 'gamey' appearance. Feet strong and well spread. Toes long, strong, straight, the back toe low and nearly flat on ground; nails well shaped.
Plumage: Short, hard and close.
Handling: Flesh firm.

Female

The general characteristics are similar to those of the male, allowing for the natural sexual differences. The tail, however, which is well venetianed but close, is carried low but somewhat higher than the male's.

Dark Indian Game

Colour

Male plumage: Head, neck, breast, underfluff, thighs, and tail black, with rich green glossy sheen or lustre, the base of the neck and tail hackles a little broken with bay or chestnut, which should be almost hidden by the body of the feathers. Shoulders and wing bows green glossy black or beetle-green, slightly broken with bay or chestnut in the centre

Jubilee Indian Game male, large

Dark Indian Game female, large

Dark Indian Game, cockerels above, hens below. These drawings form part of the Indian Game Standard.

of the feather or shaft. Tail coverts green glossy black or beetle-green slightly broken with bay or chestnut in the base of the shaft. Back feathers green glossy black or beetle-green, also touched on the fine fronds at the end of the feathers with bay or chestnut which gives the sheen so much desired. When the wing is closed there is a triangular patch of bay or chestnut formed of the secondaries, which are green glossy black or beetle-green on the inner, and bay or chestnut on the outer web, and which when closed show only the bay in a solid triangle. The primaries, ten in number, are curved and of a deep black, except for about 6.25 cm ($2\frac{1}{2}$ in) of a narrow lacing of light chestnut on the outer web.

Female plumage: The ground colour is chestnut-brown, nut-brown, or mahogany-brown. Head, hackle and throat green glossy black or beetle-green. The pointed hackle that lies under the neck feathers green glossy black, or beetle-green with a bay or chestnut centre mark; the breast commencing on the lower part of the throat, expanding into lacing on the swell of the breast, of a rich bay or chestnut, the inner or double lacing being most distinct, the belly and thighs being marked somewhat similarly and running off into a mixture of indistinct markings under the vent and swell of the thighs. The feathers of shoulders and back are somewhat smaller, enlarging towards the tail coverts and similarly marked with double lacing; the markings on wing bows and shoulders running down to the waist are most distinct of all, with the same kind of double lacing. Often in the best specimens there is an additional mark enclosing the base of the shaft of the feather and running to a point in the second or inner lacing. Tail coverts are seldom as distinctly marked but have the same style of marking. Primary or flight feathers are black, except on inner frond or web which is a little coloured or peppered with a light chestnut. Secondaries are black on the inner web, while the outer web is in keeping with the general ground colour and is edged with a delicate lacing of green glossy black or beetle-green. Wing coverts which form the bar are laced like those of the body and often a little peppered. The black lacing should be metallic-green, glossy black or beetle-green. This should appear embossed or raised.

In both sexes: Beak horn, yellow, or horn striped with yellow. Eyes from pearl to pale red. Face, comb, wattles and ear-lobes rich red. Legs rich orange or yellow, the deeper the better.

Weights

Male 3.60 kg (8 lb) min.
Female 2.70 kg (6 lb) min.

Scale of points

Type and colour (body and thighs 10, back, breast, wings, tail, legs, 8 each, neck 3)	53
Carriage	12
Size	10
Head (skull, eyes and brows, 3 each, beak, wattles, lobes, comb, 2 each)	17
Condition	8
	100

Note: The Indian Game fowl is in no way allied to the English Game fowl. Hence, it is not recognized as a true Game bird in the Fancy; that is, unless classes are specially provided for the breed it must compete in the 'Any other variety' classes and not in those set aside for 'Game', *vide* Poultry Club Show Rules.

Defects

Crooked breast or toes. Flat shins. Rusty hackles. Bad shape. Heavy feathering. White in hackles. Smallness of size. Long legs and thighs. Twisted hackle.

Disqualifications

Male: Crooked back, beak and legs. Wry or squirrel tail, in-knees, bent legs and flat sides. Single or Malay comb. Red hackles. Additionally in the female too light, too dark or mealy ground colour, and defective markings.

Jubilee Indian Game

General characteristics

As for Indian Game.

Colour

Male plumage: Head, neck, breast, body, underfluff, thighs and tail white. Hackle feathers to have chestnut shaftings. Clear breasts are desirable. Wings bows and shoulders white, slightly broken with bay or chestnut; wing primaries and secondaries white with

bay markings; triangular patch of bay or chestnut to show when wing is closed. Tail coverts white. Back white touched with bay or chestnut.
Female plumage: Ground colour chestnut-brown or mahogany. Head hackle and throat white. Breast, commencing on the lower part of the throat and expanding to double lacing on the swell of the breast, mahogany laced with white. The inner, or double lacing, to be most distinct. The underparts and thighs are marked somewhat similarly and run into a mixture of indistinct markings beneath the vent and swell of the thighs. Feathers of the shoulders and back somewhat small, enlarging towards the tail coverts similarly marked with the double lacing; often in the best specimens there is an additional mark enclosing the base of the shaft of the feather and running to a point in the second or inner lacing. The tail coverts are seldom as distinctly marked, but with the same style of marking. Wing primaries, white marked on inner web with chestnut; secondaries, white inner web, chestnut outer web, edged with white. Main tail white. Remainder chestnut ground colour throughout, double laced with white, inner lacing should be quite distinct. Underparts and thighs may be less distinctly marked and wing coverts may be peppered.

In all other respects the Indian Game standard should be followed.

In both sexes: Beak, eyes, comb and legs as described for Indians.

Double Laced Blue Indian Game

General characteristics

As Indian Game.

Colour

Male plumage: Lower throat, breast, thighs, belly fluff and tail even blue which can range from pigeon to dark blue, evenness of shade taking precedence. A slight edging of a darker shade round feathers permissible. Head, neck hackles, saddle, back and wing bows a darker shade of blue to the remainder of body, slightly broken with dark bay, this being mostly covered giving an overall impression of dark slate-blue. Undercolour pale blue-grey. Wing primaries blue to match main body colour with a fine chestnut outer edge; secondaries blue inner web, bay outer web, forming a solid bay triangle when the wing is closed; wing bar blue to match the main body.
Female plumage: Head, neck hackles and tail even dark slate-blue. Lower neck hackles may be slightly broken in the centre with dark bay, but this not showing through. The remainder of the body to be chestnut to mahogany-brown ground colour, evenness of utmost importance. Each feather is to be laced with two concentric rings of clear even blue, lighter in shade to the neck and tail. Lacings to be as crisp as possible and encircle the feather completely, must not be crescent shaped, too heavy or run into each other. A third blue marking in the centre base of the feather producing triple lacing is desirable. Lacing to cover the whole remainder of the body, including tail coverts and extending to the thighs, but becoming indistinct into the belly fluff and round the vent, where it becomes even blue. Undercolour pale blue-grey. Wing primaries even blue with a little fine chestnut peppering on inner web; secondaries inner web blue, outer web to match body ground colour, with a clear lacing of blue round the outer web; wing bar double laced to match the remainder of the body.

In both sexes: Beak yellow, horn or horn striped with yellow. Eyes pearl to pale red. Face, comb, wattle and ear-lobes rich red. Legs rich orange or yellow, nails to match beak.

Weights, scale of points and defects and disqualifications

As Dark Indians.

BANTAM

These miniatures are well established but they are not recognized as Game birds in spite of their name; and although they may compete in classes specified for 'hard feather' they may not compete in classes listed for Game birds. Where no 'hard feather' or breed classes are scheduled, they must compete in the A.O.V. section. Jubilees are similar to the Indians except that where Indians are black, Jubilees are white. Both colours are frequently interbred.

Indian, Jubilee and blue laced Indian Game bantams should follow the large fowl standard, including the scale of points.

Weights

The Indian Game Club does not issue definite weight standards for bantams and weights detailed below are suggestions only. These weights are often exceeded and *excess size should be penalized.*

Male 2.00 kg ($4\frac{1}{2}$ lb)
Female 1.50 kg ($3\frac{1}{2}$ lb)

Originally, British Bantam Association weights were below these. No characteristic of the breed must be developed or exaggerated to such an extent that Indian Game cannot effectively reproduce themselves.

IXWORTH

LARGE FOWL

Origin: Great Britain
Classification: Heavy: Rare
Egg colour: Tinted

The all-white Ixworth was created by Reginald Appleyard in 1932, taking its name from the village in Suffolk, and was produced as an excellent table bird with good laying qualities. Breeds used in its make-up included white Sussex, white Orpington, white Minorca, Jubilee, Indian and white Indian Game. In 1938, Ixworth bantams were to follow and at the time their breeder said they were better than the large fowl. The breed is now kept by the dedicated few, in both large fowl and bantams.

General characteristics: male

Carriage: Alert, active and well balanced.
Type: Body deep, well rounded, fairly long but compact. Back long, flat, reasonably broad, without too prominent a slope to the tail. Breast broad, full, deep, well rounded, long and wide, low breastbone carried well forward; with unpronounced keel or keel point; well fleshed and rounded off for entire length. Wings strong, carried close, showing shoulder butts. Tail compact, of medium length and carried fairly low, the sickles close fitting.
Head: Broad and of medium length. Beak short and stout. Eyes full, prominent, keen expression, without heavy brows. Comb pea type. Face smooth and of fine texture. Ear-lobes and wattles medium size and fine texture.
Neck: Somewhat erect and of reasonable length. Hackle feathers short, close fitting and in no way excessive or loose.
Legs and feet: Legs well apart, and of reasonable length to ensure activity. Thighs well fleshed and of medium length. Shanks covered with tight scales, free from feathers. Toes, four, straight, well spread and firm stance. Bone characteristic of a first-class table bird.
Plumage: Short, silky and close fitting; fluff likewise.

Female

The general characteristics are similar to those of the male, allowing for the natural sexual differences.

Ixworth male

Colour

Male and female: White.

In both sexes: Beak white. Eyes red or bright orange. Comb, face, wattles, and ear-lobes brilliant red. Legs, feet, skin and flesh white.

Weights

Cock	4.10 kg (9 lb);	cockerel	3.60 kg (8 lb)
Hen	3.20 kg (7 lb);	pullet	2.70 kg (6 lb)

Scale of points

Table merits	40
Shape and size	20
Colour (general)	20
Head	10
Plumage and condition	10
	100

Serious defects

Coarseness. Lack of activity. Loose feathers. Any point against table values or general usefulness. Any deformity.

BANTAM

Ixworth bantams should follow exactly the standard for large fowl.

Weights

Male	1020 g (36 oz)
Female	790 g (28 oz)

JAPANESE BANTAM

Origin: Japan
Classification: True bantam
Egg colour: White or cream

True bantams of great antiquity, these are without counterparts in large breeds. They are the shortest legged of all varieties and are standardized in three feather forms: plain or normal feather, frizzle feather and silkie feather. The frizzle feathered shall follow both the type and colours of the plain feathered standards, but the ends of all feathers are to curl back and point towards the head. Feathers must be broad and as closely curled as possible. The silkie feathered refers to feather construction. All birds must follow closely the general standard, but body feathers shall have a silky, loose feather structure (i.e. feathers have no main centre vein). This cannot apply to primary and secondary wing feathers or to true tail feathers, which would nullify any true Japanese type.

General characteristics: male

Carriage and appearance: Very small, low built, broad and cobby with deep full breast and full-feathered upright tail. Appearance somewhat quaint due to a very large comb, dwarfish character and waddling gait. Plumage very full and abundant.
Type: Back very short, wide, and seen from the side it forms the shape of a small letter U, the sides being formed by the neck and tail. This shape, however, is almost lost in fully feathered males. Saddle hackles rich and long. Body short, deep and broad. Breast very full, round and carried prominently forward. Wings long with the tips of the secondaries touching the ground immediately under the end of the body. Thighs very short and not visible. Tail very large and upright. The main tail feathers should rise above the level of the head about one-third of their length, spreading well and with long sword-shaped main sickles and numerous soft side hangers. The tail may touch the comb with its front feathers, but must not be set so as to lean forward at too sharp an angle (squirrel tailed).
Head: Large and broad, beak strong and well curved, eyes large. Comb single, large (the larger the better), coarse grained, erect and evenly serrated with four or five points. The blade of the comb should follow the nape of the neck. Face smooth, ear-lobes medium size, red and free from all traces of white. Wattles pendant and large.
Neck: Rather short, curving backwards and with abundant hackle feathers which should drape the shoulders well.
Legs and feet: Shanks very short, clean (free from feather), strong and sharply angled at the joints. The shanks to be so short as to be almost invisible. Toes, four, straight and well spread.

Female

The general characteristics should follow closely those described for the male regarding type. Breast should be all as described for the male. Tail well spread and rising well above the head. The main tail feathers broad, the foremost part being slightly curved (sword-shaped). Comb large, evenly serrated and preferably erect, although falling to one side being no defect.

Colour

The black-tailed white

Male and female plumage: Body feathers white, wing primaries and secondaries should have white outer and black inner webs, the closed wings look almost white. Main tail feathers black. Main sickles and side hangers black or black with white edges.
In both sexes: Eyes red. Legs yellow.

The black-tailed buff

Male and female plumage: The same markings as the black-tailed white, except that the white is replaced by buff.

Japanese bantam male (left)
Japanese bantam female (above)

The buff columbian

Male and female plumage: Rich even buff, wing primaries and secondaries buff with black inner webs, the closed wings look almost buff. Sickles and side hangers black with buff edges. Neck hackle feathers buff with black centre down each, the hackle to be free from black edges.

In both sexes: Eyes red. Legs yellow.

The white

Male and female plumage: Pure white without sappiness.

In both sexes: Eyes red. Legs yellow.

The black

Male and female plumage: Deep full black with a green sheen.

In both sexes: Eyes red or dark. Comb and face red. Legs yellow, black permitted on shanks but underside of feet must be yellow.

The greys

The birchen grey

Male plumage: Silver neck and saddle hackles streaked with black, breast laced to top of thighs, retaining light shaft to centre of feather. Tail black with green sheen.

Female plumage: Silver laced neck hackle to be clear cut, back and tail black. Breast feathers laced to top of thighs retaining light shaft to centre of feather. Lacing to be clear cut completely surrounding each feather.

In both sexes: Red face and eye, dark face and eye if possible. Legs yellow or dusky with yellow soles.

The silver grey

Male plumage: As birchen grey, lacing to be most pronounced.
Female plumage: Colour and lacing as in birchen grey, silvering to be allowed to extend over back, wings and thighs.

In both sexes: Red face and eye, yellow or dusky legs with yellow soles.

The dark grey

Male plumage: As birchen grey, but breast black.
Female plumage: As birchen grey female without lacing to breast, clear black breast.

The Miller's grey

Male and female plumage: As birchen grey but mealy (stippled with a lighter shade as though dusted with meal) on breast.

All greys to have black wing bays.

The mottleds

The black: Male and female plumage: All feathers should be black with white tips. The amount of white may vary, but the ideal is between 0.94 cm ($\frac{3}{8}$ in) and 1.25 cm ($\frac{1}{2}$ in). Tails and wings similar but more white permitted.
The blue: Male and female plumage: As above but blue instead of black.
The red: Male and female plumage: As the black but red instead of black.

In both sexes and all mottleds

Beaks to match legs, which should be yellow or willow. Eyes red or orange.

The blues

The self: Male and female plumage: All feathers blue, neck feathers may be a darker blue than the remainder.

In both sexes: Eyes orange. Legs slate or willow.

The lavender: Male and female plumage: All feathers a lavender-blue to skin, even shade throughout.

In both sexes: Eyes orange or dark brown. Legs slate or blue.

The cuckoo

Male and female plumage: The feathers throughout including body, wing and tail to be generally uniformly cuckoo coloured with transverse bands of dark bluish grey on a light grey ground.

In both sexes: Beak yellow marked with black. Eyes orange or red. Legs yellow.

The red

Male and female plumage: All feathers deep red, solid to skin an even shade throughout.

In both sexes: Beak and legs yellow, both may be marked with red. Eyes red.

The tri-coloured

Male and female plumage: The colours white, black and brown or dark ochre should be as equally divided as possible on each feather.

In both sexes: Eycs red or orange. Legs yellow or willow.

The black-red

Male and female plumage: Wheaten and partridge bred.

The brown-red, blue-red, silver and golden duckwing

The latter four colours all as described for Old English Game.

The following secondary colours are permitted: ginger, blue dun, honey dun, golden hackled, furness.

Weights

Male	510–600 g (18–20 oz)
Female	400–510 g (14–18 oz)

Scale of points

Type	55
Size	15
Condition	15
Colour	10
Leg colour	5
	100

Serious defects

Narrow build. Long legs. Long back. Wry tail. Tail carried low. Deformed comb and lopped comb on males. High wing carriage. White lobes. Any physical deformity.

JERSEY GIANT

LARGE FOWL

Origin: America
Classification: Heavy: Rare
Egg colour: Tinted to brown

Originated in New Jersey in about 1880, this American breed took the name of 'Giant' because of the extra heavy weights that specimens could record. Its make-up accounts for such poundage as it includes black Java, dark Brahma, black Langshan and Indian Game. When introduced into this country it was claimed for the breed that the birds were heavier than those of any breed, that it was adaptable for farm range, and also for providing capons. Earlier specimens were of exceptional weights.

General characteristics: male

Carriage: Bold, alert and well balanced.
Type: Body long, wide, deep and compact; smooth at sides, with long keel, smooth and moderately full fluff. Back rather long, broad, nearly horizontal, with a short sweep to the tail. Breast broad, deep and full, carried well forward. Wings medium sized, well folded, carried at the same angle as the body, the primaries and secondaries broad and overlapping in natural order when the wings are folded. Tail rather large, full, well spread, carried at an angle of 45° above the horizontal, the sickles just sufficiently long to cover the main tail feathers, the coverts moderately abundant and of medium length, the main tail feathers broad and overlapping.
Head: Rather large and broad. Beak short, stout and well curved. Eyes large, round, full and prominent. Comb single, straight, upright, rather large and of fine texture, having six well-defined and evenly serrated points, the blade following the shape of the neck. Face smooth and fine in texture. Ear-lobes smooth and rather large, extending down one-half of the length of the wattles. Wattles of medium size and fine texture, well rounded at the lower ends.
Neck: Moderately long, full and well arched.
Legs and feet: Legs straight and set well apart. Thighs large, strong, of moderate length and well covered with feathers. Shanks strong, stout, medium length and free from feathers; scales fine, bone of good quality and proportionate to size of bird. Toes, four, of medium length, straight and well spread.

Jersey Giant large fowl
White male
Black female

Female

With the exception of the tail, which is well spread and carried at an angle of 30° above the horizontal, the general characteristics are similar to those of the male, allowing for the natural sexual differences.

Colour

The black

Male and female plumage: The surface a lustrous green-black, and the undercolour slate or light grey.

In both sexes: Beak black, shading to yellow towards the tip. Eyes dark brown or hazel. Comb, face, wattles and ear-lobes red. Legs and feet black, with a tendency towards willow in adult birds, the underpart of the feet being yellow.

The white

Male and female plumage: The surface and undercolour white.

The blue

Male and female plumage: Modern Langshan blue laced preferred.

In both sexes and all colours

Beak willow (some yellow permissible at present). Eyes brown to black. Comb, face, wattles and ear-lobes red. Legs and feet willow, i.e. dark greenish-yellow, soles yellow. Skin nearly white.

Weights

Cock	5.90 kg (13 lb);	cockerel	5.00 kg (11 lb)
Hen	4.55 kg (10 lb);	pullet	3.60 kg (8 lb)

Scale of points

Shape and carriage	25
Colour	20
Quality	15
Head	10
Size and symmetry	10
Condition	10
Legs and feet	10
	100

Serious defects (in whites)

Smoky surface colour. Side sprigs to comb. More than 0.90 kg (2 lb) below standard weight in mature stock.

Defects (in black and whites)

Overhanging eyebrows. Sluggishness. Coarseness. Excessive or superfine bone. In blacks: black or dull black undercolour extending to the skin of the hackle, back, breast, or body and fluff. Positive white showing on surface of plumage. Other than yellow under the feet. More than 0.90 kg (2 lb) below the standard weight in mature stock.

BANTAM

Jersey Giant bantams should follow exactly the standard of the large fowl.

Weights

Male	1.74 kg ($3\frac{1}{4}$ lb)
Female	1.13 kg ($2\frac{1}{2}$ lb)

KO-SHAMO BANTAM

Origin: Japan
Classification: Hard feather: Rare
Egg colour: Cream or tinted

In Japan Ko-Shamo are the most popular of the 'Small-Shamo' breeds. Other breeds of this group are Kimpa, Nankin-Shamo, Yakido and Yamato-Gunkei. Ko-Shamo have very scanty plumage and muscular build. The aim is to have as much character as possible within the weight limits.

General characteristics: male

Type: Back medium length and broad. Widest at the shoulders, gradually narrowing from above the thighs. Back should be straight and sloping down to the tail. Saddle hackle feathers should be sparse, thin and short. Wing tips should stop at the base of the saddle hackles, not carried low or high over the back. Breast and belly wide, deep and well rounded. Plumage is scanty with a band of bare red skin down the centre. Tail short, carried down in line with back. Main tail slightly fanned. Sickles and tail coverts short and well curved forming a 'prawn' or 'lobster' tail ('ebio' in Japanese), with feather tips pointing down and inwards. There are also some birds with 'tubular' tails ('tutuo' in Japanese).
Head: Large with prominent eyebrows to suggest ferocity. Beak thick, and deep from top to bottom, short and well curved. Comb walnut, pea or chrysanthemum. All types should be small, closely set to the head. Eyes large and penetrating. Wattles very small or absent. Ear-lobes and throat skin thick and rather wrinkled with a dewlap of bare red skin. The wrinkled skin is not developed on Ko-Shamo as much as on Yamato-Gunkei.
Neck: Long, strong, curved slightly and almost erect. The bare red skin of the dewlap extends well down the front of the neck. Neck hackle feathers are thin and short hardly reaching the base of the neck.
Legs and feet: Thighs of medium length and well muscled. Legs well apart, accentuated by the general sparse plumage of body and legs. Shanks and ankles are thick with four or more straight rows of scales. Four toes, straight and well spread.
Plumage: Very hard and sparse. Bare red skin showing at wing joints on back, around back vent and along the keel.

Female

These are similar to the male, allowing for the natural sexual differences.

Colour

The black
Male and female plumage: Glossy green-black throughout.

The black-mottle
Male and female plumage: Any mixture of black and white (called 'go-stone' in Japanese).

The black-red
Male plumage: The red may be any shade from yellow to dark red.
Female plumage: Any shade of wheaten or cinnamon with or without black markings.

The blue
Male and female plumage: Any shade, laced or self.

The buff
Male and female plumage: Normally a light shade called 'orang-utan' in Japan.

Ko-Shamo male

Ko-Shamo female

The duckwings and greys
Normally with silver wheaten females.

The spangled
Any mixture of black, white and shades of red.

The cuckoo and the white
As in Game.

In both sexes and all colours
Eyes pearl or yellow, often darker in young birds. Beak yellow or horn, with dark markings on some varieties. Legs and feet yellow, dusky on darker varieties. Skin where exposed, red.

Weights

Cock	1000 g (36 oz);	cockerel	800 g (28 oz)
Hen	800 g (28 oz);	pullet	600 g (22 oz)

Scale of points

Carriage and type	30
Head and neck	20
Condition and plumage quality	20
Legs and feet	10
Eye colour	10
Legs and feet colour	5
Plumage colour	5
	100

Note: Judges are reminded that the main objective is to achieve the desired type within weight limits. A well-built Ko-Shamo's body is to be divided into three equal parts: head and neck; breast; legs.

KRAIENKÖPPE (TWENTSE)

Origin: Netherlands and Germany
Classification: Light: Rare
Egg colour: White

Kraienköppe is the German name, Twentse the Dutch name for this border area breed. The basic breed type began with crosses between Malays and local farmyard fowls in the late-nineteenth century. Later, Silver Duckwing Leghorns were introduced. Kraienköppe were first exhibited in the Netherlands in 1920 and Germany in 1925. The bantams were developed from crosses of the large breed with Malay bantams. They were first exhibited in the Netherlands in 1940, in Germany in 1955 and in Britain around 1970.

General characteristics: male

Carriage: Upright and elegant with powerful appearance.
Type: Body extended, strongly built, becoming fuller towards the rear. Back fairly long, straight, rounded at the sides with wide and abundant saddle hackle. Shoulders powerful and fairly wide. Breast wide and full. Wings long and powerful, carried closely with the tips under the saddle hackle. Tail fairly long and carried at an angle of 30–40°, with full sickles.
Head: Short, wide, arched, with a visibly prominent nape. Face free from feathers. Comb narrow walnut, in the shape of an elongated strawberry (or acorn), well set. Wattles short. Ear-lobes small. Beak short, strong, the tip bent downwards. Eyes fiery, alert, set somewhat under beetling brows.
Neck: Powerful, wide between the shoulders, of good average length, carried upright, curved slightly backwards with abundant hackle falling over the shoulders and back.
Legs and feet: Thighs powerful, prominent with smooth feathering. Shanks slender, smooth, free from feathers. Toes, four, fairly long, widely spread.
Plumage: Tight fitting.

Female

The general characteristics are similar to those of the male, allowing for the natural sexual differences. The back is carried almost horizontally. The tail is closed but not pointed. The comb is the size of a pea, very flattened. The wattles are small to the point of disappearance.

Colour

The silver

Male plumage: Head white, neck silvery white with black shaft stripe. Wing bows and back pure silvery white. Saddle silvery white with distinct shaft stripe. Wings: bays silvery white, wide black wing bars with green sheen; primaries black with narrow white outer lacing; secondaries' outer colour white, inner colour and tips black, so that the closed wing appears pure white. Breast, abdomen, thighs and hind part black, tail pure deep black with black sickles with green sheen.
Female plumage: Head silvery grey, hackle pure silvery white with black shaft stripe; back, shoulders and wings ash-grey with silvery stippling and a whitish shaft. From hackle to tail every feather should show a narrow bright silvery grey lacing. Breast salmon to salmon-red, abdomen and hind part ash-grey; tail black and greyish black.

The golden

Male plumage: As the silver, except that the silvery white is replaced by golden-red, lighter on head and neck.
Female plumage: Head and neck golden-yellow marked as the silver. Back, shoulders and wings light brown ground colour of even shade with fine black striping, peppering and stippling; yellow shaft. From hackle to tail every feather should show a narrow golden

Silver Kraienköppe male, large

Silver Kraienköppe female, large

lacing. Breast salmon to salmon-red, abdomen and hind part brownish ash-grey. Tail black with brown markings.

In both sexes and colours

Beak yellow with dark tip. Eyes yellow-red to red. Face, comb, ear-lobes and wattles red. Legs and feet bright yellow.

Weights

Male 2.50–2.95 kg ($5\frac{1}{2}$–$6\frac{1}{2}$ lb)
Female 1.80–2.50 kg (4–$5\frac{1}{2}$ lb)

Scale of points

Type	25
Colour	25
Head	20
Legs and feet	15
Condition	15
	100

Serious defects

Short or narrow body. Roach back. Upright or poorly furnished tail. Low stance. Drooping wings. Thin neck. Coarse, pointed or narrow head. Fish eyes. Fluffy plumage. Narrow sickles. Any other comb.

BANTAM

Kraienköppe bantams should follow the large fowl standard in every respect.

Weights

Male 850 g (30 oz)
Female 740 g (26 oz)

LA FLÈCHE

LARGE FOWL

Origin: France
Classification: Heavy: Rare
Egg colour: Tinted

The La Flèche is a French breed which has never been widespread in Britain. A large black breed with two vertical spikes for a comb, it is related to the Crèvecœur and in the middle of the nineteenth century was used to produce white-skinned *petit poussin* for the Paris market.

General characteristics: male

Carriage: Bold and upstanding.
Type: Body, general appearance large, powerful and rather hard. Back wide and rather long, slanting to the tail. Wings large and powerful. Breast full and prominent. Tail moderate in size.
Head: General appearance of the head long, slightly coarse and cruel. Beak large and strong with cavernous nostrils. Comb a double spike standing nearly upright with very small spikes in front. Wattles long and pendulous, ear-lobes large. Head should be quite free of crest.

La Flèche male

Neck: Long and very upright, but not backward, with as much hackle as possible.
Legs and feet: Thighs and shanks long and powerful, the latter being free of feathers; toes large and straight.

Female

The general characteristics are similar to those of the male, allowing for the natural sexual differences.

Colour

Male and female plumage: Glossy black with bright green reflections.

In both sexes: Beak black or very dark horn, comb, wattles and face deep red, ear-lobes brilliant white. Eyes bright red or black. Legs and feet very dark slate or leaden black.

Weights

Male	3.60–4.10 kg (8–9 lb)
Female	2.70–3.20 kg (6–7 lb)

Scale of points

Type and carriage	25
Head	35
Colour	15
Size	15
Legs and feet	5
Condition	5
	100

Serious defects

Presence of crest. Entirely red ear-lobes. Feathers on legs. Incorrect leg colour. Coloured feathers. Wry tail or any deformity.

BANTAM

La Flèche bantams should follow exactly the standard of the large fowl.

Weights

Male	1020 g (36 oz)
Female	790 g (28 oz)

LAKENVELDER

LARGE FOWL

Origin: Germany
Classification: Light: Rare
Egg colour: Tinted

As the name implies this is a German breed. The attention of a few dedicated English fanciers has been attracted not only by its striking appearance, but also by its utility qualities. However, good Lakenvelder plumage is as handsome as it is difficult to obtain.

General characteristics: male

Carriage: Upright, bold and sprightly.
Type: Body moderately long, fairly wide at the shoulders and narrowing slightly to the root of the tail. Full and round breast. Broad and apparently short back. Medium long wings, tucked well up, the bows and tips covered by the neck and saddle hackles. Long and full tail, the sickles carried at an angle of 45°, but avoiding 'squirrel' carriage.
Head: Skull short and fine. Beak strong and well curved. Eyes large, bright and prominent. Comb single, erect, evenly serrated, of medium size, and following the contour of the skull. Face smooth and of fine texture. Ear-lobes small and of almond shape. Wattles of medium length, well rounded at the base.
Neck: Of medium length and furnished with long hackle feathers flowing well on the shoulders.
Legs and feet: Of medium length. Thighs well apart. Shanks fine and round, free of feathers. Toes, four, strong and well spread.

Female

The general characteristics are similar to those of the male, allowing for the natural sexual differences. (*Note:* The comb is carried erect, and not drooping.)

Colour

Male and female plumage: Black and white. Neck hackle and tail solid black, free of stripes, ticks or spots. In the male the saddle hackle is white tipped with black. Remainder pure white, undercolour slate.

In both sexes: Beak dark horn. Eyes red or bright chestnut. Comb, face and wattles bright red. Ear-lobes white. Legs and feet slate-blue.

Lakenvelder female, large

Weights

Male	2.25–2.70 kg (5–6 lb)
Female	2.00 kg ($4\frac{1}{2}$ lb)

Scale of points

Colour	45
Size	20
Head	10
Type	10
Condition	10
Legs and feet	5
	100

Serious defects

Comb other than single. Feathers on shanks. Wry tail or any other deformity.

BANTAM

Lakenvelder bantams should follow the large fowl standard.

Weights

Male	680 g (24 oz)
Female	510 g (18 oz)

LEGHORN

LARGE FOWL

Origin: Mediterranean
Classification: Light: Soft feather
Egg colour: White

Italy was the original home of the Leghorn, but the first specimens of the white variety reached this country from America around 1870, and of the brown two years or so later. These early specimens weighed not more than 1.6 kg (3.5 lb) each, but our breeders started to increase the body weight of the whites by crossing the Minorca and Malay, until the birds were produced well up to the weights of the heavy breeds. In the postwar years, the utility and commercial breeders established a type of their own, and that is the one which is now favoured. In commercial circles the white Leghorn has figured prominently in the establishment of high egg-producing hybrids.

General characteristics: male

Carriage: Very sprightly and alert, but without any suggestion of stiltiness or in-kneed appearance. Well balanced.
Type: Body wide at the shoulders and narrowing slightly to root of tail. Back long and flat, sloping slightly to the tail. Breast round, full and prominent, carried well forward; breastbone straight. Wings large, tightly carried and well tucked up. Tail moderately full and carried at an angle of 45° from the line of the back; full, sweeping sickles.
Head: Well balanced with fine skull. Beak short and stout, the point clear of the front of the comb. Eyes prominent. Comb single or rose. The single of fine texture, straight, and erect, moderately large but not overgrown, coarse or beefy, deeply and evenly serrated (the spikes broad at their base), extending well beyond the back of the head and following, without touching, the line of the head, free from 'thumb marks' and side sprigs or twist at the back. The rose moderately large, firm (not overgrown so as to obstruct the sight), the leader extending straight out behind and not following the line of the head, the top covered with small coral-like points of even height and free from hollows. Face smooth, fine in texture and free from wrinkles or folds. Ear-lobes well developed and pendant, equally matched in size and shape, smooth, open and free from folds. Wattles long, thin and fine in texture.
Neck: Long, profusely covered with hackle feathers and carried upright.
Legs and feet: Legs moderately long. Shanks fine and round – flat shins objectionable – and free of feathers. Ample width between legs. Toes, four, long, straight and well spread, the back toe straight and well spread, the back toe straight out at rear. Scales small and close fitting.
Plumage: Of silky texture, free from woolliness or excessive feather.
Handling: Firm, with abundance of muscle.

Female

With the exception of the single comb rising from a firm base and falling gracefully over either side of face without obstructing the sight, and the tail, which is carried closely and not at such a high angle, the general characteristics are similar to those of the male, allowing for the natural sexual differences.

Colour

The black

Male and female plumage: Rich green-black or blue-black, the former preferred and perfectly free of any other colour.

The blue

Male and female plumage: Even medium shade of blue from head to tail, free from

Buff Leghorn male, large

Cuckoo Leghorn male, bantam

Brown Leghorn male, large

White Leghorn female, bantam

Black Leghorn female, bantam

lacing, a dark tint allowed in the hackles of the male, but no black, 'sand' or any other colour than blue, and the more even the better.

The brown

Male plumage: Head and hackle rich orange-red striped with black, crimson-red at the front of hackles below the wattles. Back, shoulder coverts and wing bow deep crimson-red or maroon. Wing coverts steel-blue with green reflections forming a broad bar across; primaries brown; secondaries deep bay on outer web (all that appears when wing is closed) and black on the inner web. Saddle rich orange-red with or without a few black stripes. Breast and underparts glossy black, quite free from brown splashes. Tail black glossed with green; any white in tail is very objectionable; tail coverts black edged with brown.

Female plumage: Hackle rich golden-yellow, broadly striped with black. Breast salmon-red, running into maroon around the head and wattles, and ash-grey at the thighs. Body colour rich brown, very closely and evenly pencilled with black, the feathers free from light shafts, and the wings free from any red tinge. Tail black, outer feathers pencilled with brown.

The buff

Male and female plumage: Any shade of buff from lemon to dark, at the one extreme avoiding washiness and at the other a red tinge; the colour to be perfectly uniform, allowing for greater lustre on the hackle feathers and wing bow of the male.

The cuckoo

Male and female plumage: Light blue or grey ground, each feather marked across with bands of dark blue or grey, the markings to be uniform; the banding shading into the ground colour not cleanly cut but sharp enough to keep the two colours distinct.

The golden duckwing

Male plumage: Neck hackle rather light yellow or straw, a few shades deeper at the front below the wattles, the longer feathers striped with black. Back deep rich gold. Saddle and saddle hackle deep gold, shading in hackle to pale gold. Shoulder coverts bright gold or orange, solid colour (an admixture of lighter feathers is very objectionable). Wing bows the same as the shoulder coverts; coverts metallic-blue (blue-violet) forming an even bar across the wing, sharp, cleanly cut and not too broad; primaries black, with white edging on the outer web; secondaries white outer web (all that appears when the wing is closed), black inner and end of feather. Breast black with green lustre. Tail black, richly glossed with green-grey fluff at the base.

Female plumage: Head grey (a brown cap is very objectionable). Hackle white, each feather sharply striped with black or dark grey (a light tinge of yellow in the ground colour admitted). Breast and undercolour bright salmon-red (this point is very important), darker on throat and shaded off to ash-grey or fawn on the underparts. Back, wings, sides and saddle dark slate-grey, finely pencilled with darker grey or black. Tail grey, slightly darker than the body colour, inside feathers dull black or dark grey.

The silver duckwing

Male plumage: Neck hackle silver-white, the long feathers striped with black. Back, saddle and saddle hackle silver-white. Shoulders and wing bow silver-white, as solid as possible (any admixture of red or rusty feathers very objectionable). Wing coverts metallic-blue (blue-violet) forming an even bar across the wing, which should be sharp and clearly cut, and not too broad; primaries black with white edging on outer parts; secondaries white outer edge (all that appears when the wing closed), black inner and end of feathers. Thighs and underparts black. Tail black richly glossed with green, grey fluff at the base.

Female plumage: Head silver-white. Hackle silver-white, each feather sharply striped with black or dark grey. Breast and underparts light salmon or fawn, darker on throat and shaded off to ash-grey on underparts. Back, wings, sides and saddle clear delicate silver-grey or French grey, without any shade of red or brown, finely pencilled with dark grey or black (purity of colour very important). Tail grey, slightly darker than the body colour with the inside feathers a dull black or dark grey.

The exchequer

Male and female plumage: Black and white evenly distributed with some white in the undercolour, the white of the surface colour in the form of a large blob as distinct from V-shaped ticking. Wings and tail to appear white and black evenly distributed.

The mottled

Male and female plumage: Black with white tips to each feather, the tips as evenly distributed as possible. Black to predominate and to have a rich green sheen.

The partridge

Male and female plumage: This colour is fully described under Wyandotte bantams and need not be repeated here.

The pyle

Male plumage: Neck hackle bright orange. Back and saddle rich maroon. Shoulders and wing bows dark red. Secondaries dark chestnut outer web (all that appears when the wing is closed) and white inner. Remainder white.

Female plumage: Neck white tinged with gold. Breast deep salmon-red shading into white thighs. Remainder white.

The white

Male and female plumage: Pure white, free from straw tinge.

In both sexes and all colours

Beak yellow or horn. Eyes red. Comb, face and wattles bright red. Ear-lobes pure opaque white (resembling white kid) or cream, the former preferred. Legs and feet yellow or orange.

Weights

Cock 3.40 kg ($7\frac{1}{2}$ lb); cockerel 2.70–2.95 kg ($6–6\frac{1}{2}$ lb)
Hen 2.50 kg ($5\frac{1}{2}$ lb); pullet 2.00–2.25 kg ($4\frac{1}{2}$–5 lb)

Scale of points

Type	25
Comb	10
Lobe	10
Eyes	5
Legs	10
Breast	5
Size	5
Colour	20
Condition	10
	100

Serious defects in all colours (for which a bird should be passed)

Single comb – Male's comb twisted or falling over, or female's erect. Ear-lobes red. Any white on face. Legs other than yellow or orange. Side sprigs on comb. Wry or squirrel tail, or any bodily deformity. *Rose comb* – Comb other than rose or such as to obstruct sight. Ear-lobes red. White in face. Wry or squirrel tail or any bodily deformity. Legs other than orange or yellow.

BANTAM

Leghorn bantams should follow the large fowl standard in all respects.

Weight

Male 1020 kg (36 oz) max.
Female 910 kg (32 oz) max.

MALAY

LARGE FOWL

Origin: Asia
Classification: Rare: Hard feather
Egg colour: Tinted

At the first poultry show in England in 1845 the Malay had its classification, and in the first British Book of Standards of 1865 descriptions were included of both the black-red and the white Malay. One of the oldest breeds, the Malay reached this country as early as 1830, and our breeders developed it, particularly in Cornwall and Devon. At the turn of the twentieth century, the Malay was the first breed to be bantamized and proved to be more popular than the large fowl. They were large in comparison with other bantams and it is not easy to reduce them further without losing the typical large breed characteristics. They should follow the large fowl standard in every respect.

General characteristics: male

Carriage: Fierce, gaunt, very erect, high in front, drooping at stern, straight at the hock, and a hard, clean, cut-up appearance from behind.

Black-red Malay male, large

Black-red Malay female, large

Type: Body wide fronted, short, and tapering; broad and square shoulders, the wing butts prominent, well up, and devoid of feathers at the point. Back short, sloping, and with convex outline, the saddle narrow and drooping. Breast deep and full, generally devoid of feathers at the point of the keel. Wings large, strong, carried high and closely to the sides. Tail of moderate length, drooping but not whipped, the sickles narrow and only slightly curved. The outline of the neck hackle, back, and tail (upper feathers) should form a succession of curves at nearly equal angles.
Head: Very broad with well-projecting or overhanging (beetle) eyebrows, giving a cruel and morose expression. Beak short, strong, and curved downwards. The profile of the skull and the beak approaches in shape a section of a circle. Eyes deep set. Comb shaped like a half-walnut, small, set well forward, and as free as possible from irregularities. Face smooth. Ear-lobes and wattles small.
Neck: Long and upright, with a slight curve, thick through from gullet to back of skull, the bare skin of the throat showing some way down the neck; the hackle full at the base of skull, but very short and scanty elsewhere.
Legs and feet: Legs long and massive, set well on the front of the body. Thighs muscular with very little feather, leaving the hocks perfectly exposed. Shanks free of feathers, beautifully scaled, flat at the hocks and gradually rounding to the spurs, which should have a downward curve. Toes, four, long, and straight, with powerful nails, the fourth (or hind) toe close to the ground.
Plumage: In all varieties is short and scanty, hard and narrow.
Handling: Firm fleshed and muscular.

Female

With the exception of the tail, which is carried slightly above the horizontal line, well 'played' as if flexible at the joint, rather short and square and neither fanned nor whipped, the general characteristics are similar to those of the male, allowing for the natural sexual differences.

Colour

The black

Male and female plumage: Glossy black all over with brilliant green and purple lustre, the green predominant, free from brassy or white feather.

The black-red

Male plumage: Hackle, saddle, back and wing bow rich red. Secondaries bright bay; flights black inner web, red outside edging. Remainder lustrous green-black.
Female plumage: Any shade of cinnamon with dark purple-tinted hackle; quite free of ticks, spangles, or pencilling, or white in tail and wings. Partridge marked and clay females with golden hackle are also allowed.

The pyle

Male plumage: Hackle, saddle, back, and wing bow rich red. Secondaries bright bay; flights white inner web, red outside edging. Remainder cream-white.
Female plumage: Hackle gold. Breast salmon. Remainder cream-white.

The spangled

Male plumage: Breast, underparts, thighs and tail an admixture of red and white. Remainder, each feather somewhat resembling tortoiseshell in the blending of red or chestnut with black, and with a bold white tip or spangle, the flight feathers and tail as tri-coloured as possible.
Female plumage: Rich dark red or chestnut boldly marked with black and white.

The white

Male and female plumage: Pure white, free from any yellow, black, or ruddy feathers.

In both sexes and all colours

Beak yellow or horn. Eyes pearl, white, yellow or daw with a green shade but the lighter the better; a red or foxy tinge very objectionable. Comb, face, throat, wattles and ear-lobes brilliant red. Legs and feet rich yellow, although in the black a slight duskiness may be overlooked.
Note: The foregoing are the principal varieties, and others are not kept or bred in sufficient numbers to warrant description. The above colours and markings are ideal, but type and quality are the most important points in the Malay fowl.

Weights

Male	5.00 kg (11 lb) approx.
Female	4.10 kg (9 lb) approx.

Scale of points

Type (shoulders 7, curves and carriage 16, reach 12)	35
Head	16
Eyes	9
Legs	10
Feathering	10
Colour	6
Tail	6
Condition	8
	100

Note: Size to be left to the discretion of the judge.

Serious defects

Any clear evidence of an alien cross. Lack of size, in large fowl, oversize in bantams. Single, spreading, or pea comb. Red eye, bow legs, knock knees, bad feet, in short, any defect not sufficiently penalized by the deduction of the maximum number of points allowed in the above scale.

BANTAM

Malay bantams should follow the large fowl standards.

Weights

Male 1190–1360 g (42–48 oz)
Female 1020–1130 g (36–40 oz)

Most specimens exceed these weights.

MARANS

LARGE FOWL

Origin: France
Classification: Heavy: Soft feather
Egg colour: Dark brown

Taking its name from the town of Marans in France, this breed has in its make-up such breeds as the Coucou de Malines, Croad Langshan, Rennes, Faverolles, barred Rock, Brakel and Gatinaise. Imported into this country round about 1929, it has developed as a dual-purpose sitting breed. Like other barred breeds the cuckoo Marans females can be mated with males of other suitable unbarred breeds to give sex-linked offspring of the white head-spot distinguishing characteristic.

General characteristics: male

Carriage: Active, compact and graceful.
Type: Body of medium length with good width and depth throughout; front broad, full and deep. Breast long, well fleshed, of good width, and without keeliness. Tail well carried, high.
Head: Refined. Beak deep and of medium size. Eyes large and prominent; pupil large and defined. Comb single, medium size, straight, erect, with five to seven serrations, and of fine texture. Face smooth. Wattles of medium size and fine texture.
Neck: Of medium length and not too profusely feathered.
Legs and feet: Legs of medium length, wide apart, and good quality bone. Thighs well fleshed, but not heavy in bone. Shanks clean and unfeathered. Toes, four, well spread and straight.
Plumage: Fairly tight and of silky texture generally.
Handling: Firm, as befits a table breed. Flesh white, and skin of fine texture.

Female

General characteristics similar to those of the male, allowing for the natural sexual differences. Table and laying qualities to be taken carefully into account jointly.

Colour

The black

Male and female plumage: Black with a beetle-green sheen.

The dark cuckoo

Male and female plumage: Cuckoo throughout, each feather marked across with bands of blue-black. A lighter shaded neck in both male and female, and also back in the

Marans female, large

Marans male, bantam

male, is permissible if definitely banded. Cuckoo throughout is the ideal, as even as possible.

The golden cuckoo

Male plumage: Hackles bluish-grey with golden and black bands, neck paler than saddle. Breast bluish-grey with black bands, pale golden shading on upper part. Thighs and fluff light bluish-grey with medium black banding. Back, shoulders and wing bows bluish-grey with rich bright golden and black bands. Wing bars bluish-grey with black bands, golden fringe permissible. Wings, primaries dark blue-grey, lightly banded; secondaries dark blue-grey, lightly banded, with slight golden fringe. Tail dark blue-grey banded with black; coverts blue-grey banded with black. General cuckoo markings.

Female plumage: Hackle medium bluish-grey with golden and black bands. Breast dark bluish-grey with black bands, pale golden shading on upper parts. Remainder dark bluish-grey with black bands. Cuckoo markings.

The silver cuckoo

Male plumage: Mainly white in neck and showing white on upper part of breast, also on top. Remainder banded throughout, with lighter ground colour than the dark cuckoo.

Female plumage: Mainly white in neck and showing white on upper part of breast. Remainder banded throughout, with lighter ground colour than the dark cuckoo.

In both sexes and all colours

Beak, white or horn. Eyes red or bright orange preferred. Comb, face, wattles and ear-lobes red. Legs and feet white.

Weights

Cock	3.60 kg (8 lb);	cockerel	3.20 kg (7 lb)
Hen	3.20 kg (7 lb);	pullet	2.70 kg (6 lb)

Scale of points

Type, carriage and table merits (to include type of breast and fleshing, also quality of flesh)	40
Size and quality	20
Colours and markings	15
Head	10
Condition	10
Legs and feet	5
	100

Serious defects

Feathered shanks. General coarseness. Lack of activity. Superfine bone. Any points against utility or reproductive values. White in lobe.

Defects (for which a bird may be passed)

Deformities, crooked breastbone, other than four toes, etc.

Defects (in blacks)

Restricted white in undercolour in both sexes. A little darkish pigmentation in white shanks.

BANTAM

Marans bantams should be true miniatures of their large fowl counterparts. No black bantam variety is standardized.

Weights

Cock	910 g (32 oz);	cockerel	790 g (28 oz)
Hen	790 g (28 oz);	pullet	680 g (24 oz)

MARSH DAISY

LARGE FOWL

Origin: Great Britain
Classification: Light: Rare
Egg colour: Tinted

The Marsh Daisy was created around the 1880s by a Mr J. Wright of Southport, using an Old English Game bantam cock crossed on to cinnamon Malay hens. A cock produced from that cross was mated to hens which were a black Hamburgh/white Leghorn cross. A white rosecombed male produced from that cross was in turn crossed back to the hens of the Hamburgh/Leghorn cross. No other blood was introduced until 1913 when a Mr C. Moore purchased some hens from Mr Wright and crossed them on to a pure Pit Game cock. Desiring to secure the white lobe and willow leg stock, it was crossed with Sicilian Buttercups. The above were the basic ingredients for what we now know as the Marsh Daisy, a moderate layer and good forager. There are no known bantams in this breed.

General characteristics: male

Carriage: Upright, bold and active.
Type: Body long, fairly broad, especially at the shoulders, with square and blocky appearance. Almost horizontal back. Well rounded and prominent breast. Full tail, carried at an angle of 45° from the vertical.
Head: Skull fine. Beak short and well curved. Eyes bold and prominent. Comb rose, medium size, well and evenly spiked, finishing in a single leader 1.25 cm ($\frac{1}{2}$ in) long in line with the surface, not as high as the Hamburgh's or following the nape of the neck as the Wyandotte's. Face smooth. Ear-lobes almond shaped. Wattles of fine texture and in keeping with the comb.
Neck: Fairly long, fine. Hackle flowing and falling well on the shoulders to form the cape.
Legs: Moderately long. Shanks and feet light boned, free from feathers. Toes, four, well spread.
Plumage: Semi-hard, of fine texture; profuse feathering to be deprecated.

Female

The general characteristics are similar to those of the male, allowing for the natural sexual differences.

Colour

The black

Male and female plumage: Black, with beetle-green sheen in abundance.

The buff

Male and female plumage: Golden-buff throughout and buff to the skin. (*Note:* The male's tail is black to bronze but the ideal is a whole buff bird.)

The brown

Male plumage: Neck hackle rich gold, back and saddle dark gold. Main tail black; sickles black; coverts black, the whole to have a beetle-green sheen. Saddle hackle dark

gold, a little lighter gold at tips not objectionable. Wing bow dark gold, same shade as back; coverts or bar, black with beetle-green sheen; secondaries forming the bay a flat brown, showing a triangular brown bay; primaries a flat black, with the lower edge flat brown, and all well hidden when the wing is closed and tucked up. Breast and all underbody parts black with patches of golden-brown spangled in; solid, shiny black should be striven for in these parts. Undercolour decided blue to blue-grey, with a little buff or light golden-brown in places on breast.
Female plumage: Head and hackle rich gold, the tips of all feathers black, the whole to form a fringe at the cape. Back and wings brown ground ticked or peppered with darker brown or flat black; this may result in a series of black bars across the feathers, which is not objectionable. Tail dull, flat black, a little lighter at the edge of the feathers not a disqualification., but should be discouraged. Breast and all underbody parts red-wheaten or salmon, a level shade all over: too light a shade for these parts, or too deep a red-wheaten, should not be striven for.

The wheaten

Male plumage: Hackles rich gold. Back and wing bow deep gold. Tail (coverts and sickles) rich beetle green-black. Remainder golden-brown, the colour of a fairly dark bay horse. Undercolour (seen when the feathers are raised) from smoke white to a French or blue-grey, a little light buff fluff at the skin of the breast permissible.
Female plumage: Hackle chestnut with black tips forming a fringe at the base of it. Shoulders and back (upper part) red-wheat; lower part of back to root of tail lighter shade, due to the feathers having a white-wheat edging and red-wheat centre and giving a dappling effect. Wing bows red-wheat, the flights presenting a triangular patch of light brown when closed. Breast white-wheat. Tail dull black with red-wheat edging. Undercolour of back, smoke white to blue-grey; of breast, pure white.

The white

Male and female plumage: Pure white.

In both sexes and all colours

Beak horn. Eyes rich red with black pupil. Comb, face and wattles red. Ear-lobes white. Legs and feet pale willow green; toenails horn.

Weights

Male	2.50–2.95 kg ($5\frac{1}{2}$–$6\frac{1}{2}$ lb)
Female	2.00–2.50 kg ($4\frac{1}{2}$–$5\frac{1}{2}$ lb)

Scale of points

Head (lobes 13, comb and wattles 10, other points 10)	33
Plumage	20
Condition	20
Type	15
Legs	12
	100

plus 'Laying power', 20

Serious defects

Want of type. Less than one-third white lobe. Red plumage, legs other than willow green.

MINORCA

LARGE FOWL

Origin: Mediterranean
Classification: Light: Soft feather
Egg colour: White

The Minorca has been developed in this country as our heaviest light breed, and was at one time famous for its extra-large, white eggs. Crossing with the Langshan and other heavy breeds did not improve the egg production of the breed, and concentration on exaggerated headgear had a similar effect. Those times are passed and wiser counsels now prevail. The result is that a much better, balanced type is aimed for on the show-bench with moderate size of lobes and of comb, and a more prominent frontal.

General characteristics: male

Carriage: Upright and graceful.
Type: Body broad at the shoulders, square and compact. Breast full and rounded. Wings moderate in length, neat fitting close to body. Tail full, sickles long, well arched and carried well back.
Head: Long and broad, so as correctly to carry the comb quite erect. Beak fairly long and stout. Eyes full, bright and expressive. Comb single or rose in large fowl, single only in bantams. The single large, evenly serrated, five serrations preferred, perfectly upright, firmly set on the head, straight in front, free from any twist or thumb mark, reaching well to the back of the head, moderately rough in texture and free from any side sprigs. The rose oblong shape, broad at the base over the eyes, closely fitting, upright, firmly carried, full in front and tapering gradually to the 'leader' at the back, surface evenly covered with small nodules or points, free from hollowness, the 'leader' to follow the curve of the neck but not to touch the hackle. Face fine in quality, as free from feathers or hairs as possible, and not showing any white. Ear-lobes medium in size, almond shaped, smooth, flat, fitting close to the head. The lobe should not exceed 6.88 cm ($2\frac{3}{4}$ in) deep and 3.75 cm ($1\frac{1}{2}$ in) at its widest point on the top, tapering as the Valencia almond in shape. No definite size for lobes is fixed in bantams. Wattles, long and rounded at the end.
Neck: Long, nicely arched, with flowing hackle.
Legs and feet: Legs of medium length, and thighs stout. Toes four.

Female

The general characteristics are similar to those of the male, allowing for the natural sexual differences, with the exception of the single comb which drops well down over the side of the face, so as not to obstruct the sight, and the ear-lobes which are 4.44 cm ($1\frac{3}{4}$ in) deep and 3.13 cm ($1\frac{1}{4}$ in) wide.

Colour

The black

Male and female plumage: Glossy black.

In both sexes: Beak dark horn. Eyes dark. Comb, face and wattles blood red. Ear-lobes pure white. Legs black or very dark slate, the latter in adults only.

The white

Male and female plumage: Glossy white.

In both sexes: Beak white. Eyes red. Comb, face and wattles blood red. Legs pinky white.

The blue

Male and female plumage: Soft medium blue, free from lacing, the male darker in hackles, wing bows and back.

Minorca male, bantam

Minorca female, bantam

In both sexes: Beak, comb, face, ear-lobes and wattles as in the black. Eyes dark brown (darker the better). Legs blue to slate.

Weights

Cock	3.20–3.60 kg (7–8 lb);	cockerel	2.70–3.60 kg (6–8 lb)
Hen	2.70–3.60 kg (6–8 lb);	pullet	2.70–3.20 kg (6–7 lb)

Scale of points

The black and white

Style, symmetry (type)	10
Size	15
Face	15
Comb	15
Ear-lobes	10
Legs, eyes and beak	8
Colour	10
Condition	10
Breastbone	7
	100

The blue

Type	15
Size	10
Condition	10
Head, comb and ear-lobes	5
Colour	60
	100

Serious defects

White or blue in face. Feathers on legs. Other than four toes. Wry or squirrel tail. Plumage other than black, blue or white in the several varieties. Legs other than black or slate in the black, blue or slate-blue in the blue, or white in the white. Side sprigs in comb.

BANTAM

Minorca bantams should be miniatures of their large fowl counterparts and standard points, colour and defects to be the same for bantams as for large fowl.

Weights

Male	960 g (34 oz)
Female	850 g (30 oz)

Smaller specimens to be favoured, other points being equal.

MODERN GAME

LARGE FOWL

Origin: Great Britain
Classification: Hard feather
Egg colour: Tinted

By the introduction of Malay crosses, and with the skill of British fanciers, the Modern Game fowl was evolved. Black-red, duckwings, brown-reds, piles, and birchens were the

Silver duckwing Modern Game female, large

Pile Modern Game male, bantam

original recognized varieties, the general characteristics being the same for each, and 13 colours are now standardized.

General characteristics: male

Carriage: Upstanding and active. In the show-pen the bird should show plenty of 'lift' as if reaching to its fullest height.
Type: Body short, flat back, wide front and tapering to the tail, shaped like a smoothing iron. Shoulders prominent and carried well up. Wings short and strong. Tail short, fine, closely whipped together, carried slightly above the level of the body, the sickles narrow, well pointed and only slightly curved.
Head: Long, snaky and narrow between the eyes. Beak long, gracefully curved and strong at the base. Eyes prominent. Comb single, small, upright, of fine texture, evenly serrated. Face smooth. Ear-lobes and wattles fine and small to match the comb.
Neck: Long and slightly arched, fitted with 'wiry' feathers, but thin at the junction with the body.
Legs and feet: Legs long and well rounded. Thighs muscular. Shanks free of feathers. Toes, four, long, fine and straight, the fourth (or hind) toe straight out and flat on the ground, not downwards against the ball of the foot (or 'duck-footed'), which is most objectionable.
Plumage: Short and hard.

Female

The general characteristics are similar to those of the male, allowing for the natural sexual differences.

Colour

Colour is very important in Modern Game and carries 20 points. Varieties include 13 standardized colours, the blues being included in 1985 and the wheaten in 1993.

Legs and beaks vary with the colour-varieties, from yellow in piles and whites through willow in black-reds to black in birchens. Eyes similarly vary from bright red to black.

Combs and faces vary from bright red to dark purple and black. Leg colours are definitely 'tied' to each variety. Thus, whites and pyles must always have yellow legs, while shanks are willow in duckwings and black in brown-reds.

The birchen

Male plumage: Hackle, back, saddle, shoulder coverts and wing bows silver-white, the neck hackle with narrow black striping. Remainder rich black, the breast having a narrow silver margin around each feather, giving it a regular laced appearance gradually diminishing to perfect black thighs.
Female plumage: Hackle similar to that of the male. Remainder rich black, the breast very delicately laced as in the male.

In both sexes: Beak dark horn. Eyes black. Comb, face, wattles and ear-lobes dark purple. Legs and feet black.

The black-red

Male plumage: Cap orange-red. Neck hackle light orange, free from black stripe. Back saddle rich crimson. Wing bow orange; bar green-black; primaries black; secondaries rich bay on the outer edge, black on the inner and tips, the rich bay alone showing when the wing is closed. Remainder green-black.
Female plumage: Hackle gold, slightly striped with black, running to clear gold on the cap. Breast rich salmon, running to ash on thighs. Tail black, except the top feathers, which should match the body colour. Remainder light partridge-brown, very finely pencilled, and a slight golden tinge pervading the whole, which should be even throughout, free from any ruddiness whatever and with no trace of pencilling on the flight feathers.

In both sexes: Beak dark green. Eyes, comb, face, wattles and ear-lobes bright-red. Legs willow.

The brown-red

Male plumage: Hackle, back and wing bow bright lemon, the neck hackle feathers striped down the centre with green-black, not brown. Remainder green-black, the breast feathers edged with pale lemon as low as the top of the thighs.
Female plumage: Neck hackle light lemon to the top of the head, the lower feathers being striped with green-black. Remainder green-black, the breast laced as in the male, the shoulders free from ticking and the back from lacing. (*Note:* There should be only two colours in brown-red Game, viz. lemon and black. In the male the lemon should be very rich and bright, and in the female light; the black in both sexes should have a bright green gloss known as beetle-green.)

In both sexes: Beak very dark horn, black preferred. Eyes, comb, face, wattles, ear-lobes, legs and feet black.

The golden duckwing

Male plumage: Hackle cream-white, free from striping. Back and saddle pale orange or rich yellow. Wings: bow pale orange or rich yellow; bars and primaries black with blue sheen; secondaries pure white on the outer edge, black on inner and tips, the pure white alone showing when the wing is closed. Remainder black with blue sheen.
Female plumage: Hackle silver-white, finely striped with black. Breast pale salmon, diminishing to ash-grey on thighs. Tail black, except top feathers, which should match the body colour. Remainder French or steel-grey, very lightly pencilled, and even throughout.

In both sexes: Beak dark horn. Eyes ruby-red. Comb, face, wattles and ear-lobes red. Legs and feet willow.

The silver duckwing

Male plumage: Hackle, back, saddle, shoulder coverts and wing bows silver-white. Secondaries pure white on the outer edge and black on the inner, with tips of bay, the white alone showing when the wing is closed. Remainder lustrous blue-black.

Female plumage: Hackle silver-white, finely striped with black. Breast pale salmon, diminishing to pale ash-grey on thighs. Tail black, except top feathers which should match the body colour. Remainder light French grey with almost invisible black pencilling.

In both sexes: Beak, etc. as in the golden duckwing.

The pyle

Male plumage: Hackle one shade of bright orange-yellow. (Dark or washy hackles to be avoided.) Back and saddle rich maroon. Wing bow maroon; bar white and free from splashes; primaries white; secondaries dark chestnut on the outer edge and white on the inner and tips, the dark chestnut alone showing when wings is closed. Remainder pure white.

Female plumage: Hackle white, tinged with gold. Breast rich salmon-red. Remainder pure white.

In both sexes: Beak yellow. Eyes bright cherry-red. Comb, face, wattles and ear-lobes red. Legs and feet rich orange-yellow.

The wheaten

Male plumage: Cap orange-red. Neck hackle light orange, free from black striping. Back and saddle orange to crimson. Wing bow orange; bar green-black; primaries black; secondaries rich bay on outer edge, black on inner and tips, the rich bay alone showing when the wing is closed. Remainder green-black. Altogether showing a brighter top colour than the 'partridge' black-red male.

Female plumage: Hackle golden-orange very lightly striped with black running clear on the cap. Breast light salmon diminishing to fawn or cream on thighs. Body colour pale cinnamon or wheaten. Primaries black with wheaten to outer edge; secondaries black with dark wheaten to outer edge. Tail black except top feathers which, with tail coverts, are darker wheaten.

In both sexes: Beak yellowish horn. Eyes, comb, face, wattles and ear-lobes red. Legs willow, often showing yellowish soles, a distinguishing mark of the wheaten.

The black

Male and female plumage: Black, free from any other colour, with a purple or green sheen (the latter preferred).

In both sexes: Beak and eyes black. Comb, face, wattles and ear-lobes dark. Legs and feet black.

The blue

Male and female plumage: An even, clear, rich medium to pale blue free from lacing but with darker blue top colour in males and hackles of females.

In both sexes: Beak black or blue. Eyes dark. Comb, face, wattles and ear-lobes mulberry. Legs and feet black or blue.

The white

Male and female plumage: Pure white, free from any other colour.

In both sexes: Beak yellow. Eyes, comb, face, wattles and ear-lobes red. Legs and feet yellow.

The blue-red

Male plumage: Cap orange, hackles and saddle gold, free from blue striping. Back and wing bow rich red; bars lustrous blue; outer edge secondaries and edging of lower primaries bay, the bay alone showing when the wing is closed. All other sections an even, clear, rich medium to pale blue except tail which is a darker shade of blue.

The silver blue

Male plumage: Cap, hackles, back and wing bows silvery white, hackle feathers striped with blue. Remainder an even, clear, rich, medium to pale blue, breast feathers finely laced with silvery white as far down as top of thighs, with no silver shafting. Tail a darker shade of blue.

The lemon blue

Male plumage: Cap, hackles, back and wing bows bright lemon, hackle feathers striped with blue. Remainder an even, clear, rich, medium to pale blue, breast feathers finely laced with lemon as far down as top of thighs, with no lemon shafting. Tail a darker shade of blue.
Female plumage: Neck hackle light lemon to top of head, lower feathers striped with blue. Remainder an even, clear, rich medium to pale blue, breast laced as in male.
In both sexes: Beak, etc. as in the blue.

Weights

Male	3.20–4.10 kg (7–9 lb)
Female	2.25–3.20 kg (5–7 lb)

Scale of points

Type and style	30
Colour	20
Head and neck	10
Eyes	10
Tail	10
Legs and feet	10
Condition and shortness of feather	10
	100

Serious defects

Eyes other than standard colour. Flat shins. Crooked breast. Twisted toes or 'duck' feet. Wry tail. Crooked back.

BANTAM

Modern Game bantams follow the standard for large fowl. Fine body, 'reachiness' and colour are the main points in bantams. The breed is the favourite of 'die-hard' showmen.

Weights

Male	570–620 g (20–22 oz)
Female	450–510 g (16–18 oz)

MODERN LANGSHAN

LARGE FOWL

Origin: Asia
Classification: Heavy: Rare
Egg colour: Brown

The Modern Langshan has been developed on different lines from the Croad, and is quite another type of bird. It is much longer in the shanks and of a totally different outline from the original Croad.

General characteristics: male

Carriage: Graceful, upright, alert, strong on the leg with the bearing of an active bird.
Type: Body long and broad but by no means deep. Back horizontal when in normal

attitude, and with close compact plumage. Shoulders broad and abundantly furnished saddle. Wings large, closely carried, but neither 'clipped' nor 'pinched in'. Tail full, flowing, spread at base, carried fairly high but not squirrel, furnished with abundant side hangers and two sickles, each feather tapering to a point.
Head: Fine. Beak fairly long and slightly curved. Eyes large. Comb single, straight, upright, fairly small, and evenly serrated with five or six spikes. Face of fine texture. Ear-lobes medium size, pendant and inclined to fold. Wattles medium length, fine texture, neatly rounded.
Neck: Fairly long, broad at base and covered with full hackle.
Legs and feet: Legs rather long, strong and wide apart. Thighs covered with closely fitting feathers, especially around the hocks. Shanks strong but not coarse boned, with an even fringe of feathers (not heavy) on the outer sides. Toes, four, long, straight and well spread, the outer (and that alone) slightly feathered.
Plumage: Close and smooth.

Female

With the exception of the tail (not carried high) the general characteristics are similar to those of the male, allowing for the natural sexual differences.

Colour

The black

Male and female plumage: Black with a brilliant beetle-green sheen.

In both sexes: Beak dark horn to black. Eyes dark brown to black, the darker the better. Comb, face, wattles and ear-lobes brilliant red. Legs and feet dark grey with black scales in front and down the toes, showing pink between the scales (especially down the outer sides of the shanks) and on the skin between the toes. Toenails white. Underfoot pink-white. Skin of the body and thighs white and transparent.

The blue

Male plumage: Hackles, back, tail, sickles, side hangers, and wing bow rich deep slate, the darker the better, with brilliant purple sheen. Remainder clear slate-blue, each feather distinctly laced (edged) with the same dark shade as the back, the contrast between delicate ground and dark lacing being well defined.
Female plumage: Head and upper part of neck rich dark slate. Remainder clear slate-blue, each feather distinctly laced (edged) with dark slate, the lacing being well defined.

In both sexes: Beak, etc. as in the black.

The white

Male and female plumage: Pure white, with brilliant silver gloss.

In both sexes: Beak white with a pink shade near the lower edges. Legs and feet light grey or slate showing pink between the scales and on the skin between the toes. Eyes, comb, face, wattles, ear-lobes, toenails, underfoot and skin as in the black.

Weights

Cock	4.55 kg (10 lb);	cockerel	3.60 kg (8 lb)
Hen	3.60 kg (8 lb);	pullet	2.70 kg (6 lb)

Scale of points

Type and carriage	35
Head	15
Legs and feet	10
Size	10
Colour and plumage	20
Condition	10
	100

Serious defects

Yellow skin. Yellow base of beak. Yellow or orange-coloured eyes. Yellow around the eyes. Yellow shanks or underfoot. Legs other than standard colour. Shanks not feathered. More than four toes. Permanent white in the face or ear-lobes. Comb with side sprigs, or other than single. Wry or squirrel tail. Coloured feathers.

Defects

Absence of pink between toes. Feathering on middle toes. Outer toe not feathered. Too scantily or too heavily feathered shanks or outer toes. Twisted toes. Short shanks. Crooked breast. Twisted or falling comb. General coarseness. Too much fluff. Purple sheen in black; yellow shade in white.

BANTAM

Modern Langshans should follow the large fowl standard.

Weights

Male	1133 g (40 oz)
Female	910 g (32 oz)

NANKIN BANTAM

Origin: Asia
Classification: True bantam: Rare
Egg colour: Tinted

Nankins, or common yellow bantams, were among the first varieties of bantams introduced into this country. The variety came originally from Java and some parts of India. Once, they were the most widespread of all bantams and are believed to be the progenitors of nearly all buff bantam varieties. The name is thought to have been given from the resemblance of the colour to nankeen cloth. Nankins are excellent layers and the most tameable and engaging of breeds.

General characteristics: male

Carriage: Jaunty and active, with a proud bearing.
Type: Body small and neat, the breast carried well up and forward. Back short and sloping to the tail. Wings large, closely folded and carried very low, almost down to the ground. Tail large and well spread with long well-curved sickles. The whole tail carried high but not squirrel, well furnished with flowing side hangers and coverts.
Head: Small and fine. Beak rather fine and slightly curved. Eyes large and bright. Comb single or rose. The single should be neat, bearing three to six fine spikes. It should be straight and upright and in proportion with the head, carried well away from the head, a flyaway comb being characteristic. The rose comb should be small and close fitting, finely worked, with a small leader curved gracefully upward. Face smooth, ear-lobes very small, wattles small and well rounded, fine, smooth and free from wrinkles.
Neck: Medium long, well-curved and bearing long and abundant hackle.
Legs and feet: Legs short, thighs set well apart, shanks rather short, rounded, and fine, free from feathers. Toes four, rather small, straight and well spread.

Female

With the exception of the carriage, which is lower and less arrogant and with the tail carried well spread but slightly lower, the general characteristics are similar to those of the male, allowing for the natural sexual differences.

Colour

Male plumage: Back, neck and saddle hackles a rich orange, wing bows chestnut. Tail main feathers black, sickles and side hangers bronze or copper, shading into black. Remainder of plumage ginger-buff throughout to the skin.
Female plumage: Neck, back, wings, tail and saddle dark ginger-buff. Tail shading into black at the ends. Remainder of plumage light ginger-buff throughout to the skin.
In both sexes: Beak white or horn. Eyes bright orange. Comb, face, ear-lobes and wattles bright-red. Legs blue or bluish-white, male often with pinkish stripe on outside of shank. A degree of black on inner web of wings, a smaller amount or none on the female.

Weights

Male 680–737 g (24–26 oz)
Female 570–620 g (20–22 oz)

Scale of points

Type and carriage	25
Colour	20
Tail	10
Feet and legs	10
Comb	10
Other head points	10
Size	10
Condition	5
	100

Serious defects

Wry or squirrel tail. Comb other than rose or single. Comb lopped or twisted or curved down at the rear. Long legs, large feet or twisted toes. White in the ear-lobes or face. Yellow legs. Even colour throughout. White in the base of the tail. All black or buff tails in males, tail lacking black end in females. Visible black in the closed wing. Black striping or flecking on saddle and neck hackle, mealiness in surface, or grey undercolour.

NANKIN-SHAMO BANTAM

Origin: Japan
Classification: Hard feather: Rare
Egg colour: Cream or tinted

This is one of the Japanese 'Small-Shamo' breeds. Nankin-Shamo have more plumage than Ko-Shamo and do not have the wrinkled skin on the head. A regional variant exists, the Echigo-Nankin-Shamo from the Niigata area. These are very similar to the main Nankin-Shamo breed but slightly taller and slimmer.

General characteristics: male

Carriage: Body carried rather upright. Tail is rather long, straight and pheasant-like, below the horizontal.
Type: Back is long, broadest at the shoulders, sloping down to the tail. Wings are rather long, prominently away from the body at the shoulders. The flights point downwards to just behind the thighs. Saddle hackle covers the base of the tail.
Head: Medium size, showing some brow bone. Beak short, strong and curved. Eyes bold and set in oval eyelids. Comb small and compact, pea or walnut. Ear-lobes and wattle small but with some dewlap.

Nankin-Shamo female

Neck: Rather long, slightly curved, almost erect. Hackles rather long and thick but not covering shoulders.
Legs and feet: Legs are moderately long. Thighs muscular. Shanks smooth and round. Four toes, long and straight.
Plumage: Generally long feathers, but narrow and carried close to the body.

Female

As in the male, allowing for the natural sexual differences.

Colour

The black-red
Male plumage: Any shade of red.
Female plumage: Any shade from wheaten to brown, with or without dark markings.

The silver duckwings and greys
Males may be any shade. Females are usually silver-wheaten.

The spangled
Any mixture of black, white and shades of red/brown.

The black, the white
As in Game.

In both sexes and all colours
Eyes pearl or yellow. Beak yellow or horn with darker markings on black variety. Legs and feet yellow.

Weights

Cock	937 g (33 oz);	cockerel	750 g (26 oz)
Hen	750 g (26 oz);	pullet	560 g (20 oz)

Scale of points

Carriage and type	30
Head and neck	20
Condition and plumage quality	20
Legs and feet	10
Eye colour	10
Legs and feet colour	5
Plumage colour	5
	100

NEW HAMPSHIRE RED

LARGE FOWL

Origin: America
Classifications: Heavy: Rare
Egg colour: Tinted to brown

As if to copy the farmers in the State of Rhode Island who developed the breed carrying its name, those in the neighbouring State of New Hampshire developed and named their breed. The New Hampshire Red was bred by selection from the Rhode Island Red without the introduction of any other breed, taking some thirty years to reach standardization in 1935. Early maturity, quick feathering, and a plump carcase are particular features of the breed. Its body shape and colouring are very different from those of the Rhode Island Red.

General characteristics: male

Carriage: Active and well balanced.
Type: Body of medium length, relatively broad, deep and well rounded. Back of medium length, broad for entire length, gradual concave sweep to tail. Breast deep, full, broad and well rounded, the keel relatively long and extending well to the front at the breast. Wings moderately large, well folded, carried horizontally and close to the body, fronts well covered by breast feathers; primaries and secondaries broad and overlapped in natural order when wing is folded. Tail of medium length, well spread and carried at angle of 45°. Sickles medium on length, extending well beyond the main tail; lesser sickles and coverts medium length and broad; main tail feathers broad and overlapping.
Head: Of medium length, fairly deep, inclined to be flat on top rather than round. Beak strong, medium length, regularly curved. Eyes large, full and prominent, moderately high in the head. Comb single, medium size, well developed, set firmly on head, perfectly straight and upright, having five well defined points, those in front and rear smaller than these in centre; the blade smooth and inclining slightly downward, not following too closely the shape of the neck. Face smooth, full in front of the eyes; skin of fine texture. Wattles moderately large, uniform, free from folds or wrinkles. Ear-lobes elongated oval, smooth and close to head.
Neck: Of medium length, well arched, hackle abundant, flowing well over shoulders. Moderately close feathered.
Legs and feet: Legs well apart, straight when viewed from the front. Lower thighs large, muscular and of medium length. Toes, four, medium length, straight and well spread.

New Hampshire Red male, large

New Hampshire Red female, large

Plumage: Feather character to be of broad firm structure, overlapping well and fitting tightly to the body. Fluff moderately full.

Female

The general characteristics are similar to those of the male, allowing for the natural sexual differences. Comb slightly tilted at rear. Wattles of medium size, well developed and well rounded. Tail moderately well spread, carried at angle of 35°. Wings rather large carried nearly horizontal.

Colour

Male plumage: Head brilliant reddish-bay. Breast and neck medium chestnut-red. Back brilliant deep chestnut red. Saddle rich brilliant reddish-bay, slightly darker than neck. Wing fronts medium chestnut-red; bows brilliant chestnut-red; coverts deep chestnut-red; primaries, upper web medium red, lower web black edged with medium red, primary coverts black edged with medium red; secondaries, upper web medium chestnut-red having broad stripe extending along shaft to within inch of tip, lower web medium chestnut-red; shaft red. Tail, main feathers black; sickles rich lustrous greenish-black; coverts lustrous greenish-black edged with deep chestnut-red; lesser coverts deep chestnut-red. Body and fluff medium chestnut-red. Lower thighs medium chestnut-red. Undercolour in all sections light salmon, a slight smoky tinge not a defect.
Female plumage: Head medium chestnut-red. Neck medium chestnut-red, each feather edged with brilliant chestnut-red; lower neck feathers distinctly tipped with black; feathers in front of neck medium chestnut-red. Wing fronts, bows and coverts medium chestnut-red; primaries, upper web medium red, lower web medium red with narrow stripe of black extending along shaft; shaft medium red; primary coverts black edged with medium red; secondaries, lower web medium chestnut-red, upper web medium chestnut-red with black marking extending along edge of shaft two-thirds its length. Back, breast, lower thighs, body and fluff medium chestnut-red. Tail, main feathers black edged with medium chestnut-red; shaft medium chestnut-red. Undercolour as in male.

In both sexes: Beak reddish-brown. Eyes bay. Comb, face, wattles and ear-lobes bright red. Legs and toes rich yellow, tinged with reddish-horn. Line of reddish pigment down sides of shanks extending to tips of toes desirable in male.

Weights

Cock	3.85 kg ($8\frac{1}{2}$ lb);	cockerel	3.40 kg ($7\frac{1}{2}$ lb)
Hen	2.95 kg ($6\frac{1}{2}$ lb);	pullet	2.50 kg ($5\frac{1}{2}$ lb)

Scale of points

Type and carriage	25
Colour	20
Dual-purpose quality	15
Head	10
Size and symmetry	10
Legs and fect	10
Condition	10
	100

BANTAM

New Hampshire Red bantams should follow exactly the large fowl standard.

Weights

Male	980 g (34 oz)
Female	737 g (26 oz)

NORFOLK GREY

LARGE FOWL

Origin: Great Britain
Classification: Heavy: Rare
Egg colour: Tinted

The Norfolk Grey was first introduced by Mr Myhill of Norwich under the ugly name of Black Marias. They were first shown at the 1920 Dairy Show and were mainly the result of a cross breed between silver birchen Game and duckwing Leghorns. They appear regularly at shows and are plentiful in their county of origin.

General characteristics: male

Carriage: Fairly upright and very active.
Type: Body rather long, broad at shoulders. Full, round breast carried upwards. Large wings well tucked up. Well-feathered tail.
Head: Skull fine. Beak short and well curved. Eyes large and bold. Comb single, upright, of medium size, well serrated and with a firm base. Face smooth and fine. Ear-lobes small and oval. Wattles long and fine.
Neck: Of medium length, abundantly covered with hackle.
Legs and feet: Fairly short and set well back. Shanks free from feathers. Toes, four, well spread.
Plumage: Close.

Female

The general characteristics are similar to those of the male, allowing for the natural sexual differences.

Colour

Male plumage: Neck, back, saddle, shoulder coverts and wing bars silver-white, the hackles with black striping, free from smuttiness. Remainder a solid black.
Female plumage: Hackle similar to that of the male. Remainder black, the throat very delicately laced with silver (about 5 cm (2 in) only).

In both sexes: Beak horn. Eyes dark. Comb, face, ear-lobes and wattles red. Legs and feet black or slate-black, the former preferred.

Weights

Male 3.20–3.60 kg (7–8 lb)
Female 2.25–2.70 kg (5–6 lb)

Scale of points

	Male	*Female*
Colour and markings: hackles	20	15
Colour and markings: back and wings	10	15
Colour and markings: breast, thighs and fluff	15	15
Type and size	20	20
Head (comb and lobes 10, eyes 5)	15	15
Legs and feet	5	5
Condition	10	10
Tail	5	5
	100	100

Serious defects

White in lobes. Comb other than single or obstructing the sight. Legs other than black or slate-black. Feathers on shanks or feet. Lacing or shaftiness on back, breast, or wings of females.

BANTAM

Norfolk Grey bantams should follow exactly the large fowl standard.

Weights

Male 900 g (32 oz)
Female 680 g (24 oz)

NORTH HOLLAND BLUE

LARGE FOWL

Origin: Holland
Classification: Heavy: Rare
Egg colour: Tinted

The blood of the Malines in the make-up of this breed is seen in its quick maturity and rapid growth. To further its commercial importance its utility properties are valued on the show-bench in preference to markings. Lightly feathered shanks are a standardized characteristic, and the male is lighter in colour than the female. As a barred breed, the females, when mated with unbarred males of breeds with dark downs, produce sex-linked offspring, the male chicks having the white head-spot when hatched, which is absent in the female chicks.

General characteristics: male

Carriage: Upright, bold and alert.
Type: Body substantial in build, yet compact. Back broad, flat, horizontal, reasonably long, with slight rise to tail, broad saddle, prominent shoulders. Breast full, rounded and prominent; breastbone long, well fleshed and rounded off, neither shallow nor keely. Well-rounded sides, good depth of body, with a well-developed abdomen. Wings strong, well developed and close to the body. Tail broad, short, well spread with medium furnishings.
Head: Rounded and of medium length. Beak stout and short. Eyes full, bold, with keen expression, the pupils well formed and large. Face smooth, full, fine texture and without heavy eyebrows. Comb single, upright, of medium size and fine texture, with five to seven neat, even serrations, following slightly the curve of the neck at the back. Ear-lobes of medium size and silky. Wattles medium in length and size, and fine in texture.
Neck: Somewhat broad, medium in length and not too profusely feathered.
Legs and feet: Legs of medium length, wide apart, well formed and good quality bone with close scales. Thighs well developed and fleshed. Toes, four, straight, and well spread. Shanks lightly feathered, including outer toe.
Plumage: In general not too profuse. Fine in texture.
Handling: Firm, as befits a table breed. Skin thin and of fine texture. Flesh high-quality table grade.

Female

With the exception of the shanks, which are more heavily feathered, the general characteristics are similar to those of the male, allowing for natural sexual differences.

North Holland Blue female

Colour

Male plumage: Lighter blue-grey than female, barred. Undercolour very pale blue-grey, barring immaterial.

Female plumage: Dark grey-blue, slightly barred. Undercolour paler blue-grey, barring immaterial.

In both sexes: Beak white superimposed blue. Eyes orange to red. Comb, wattles, face and ear-lobes red. Shanks white, but may be shaded with blue. Skin white.

Weights

Cock	3.85–4.80 kg ($8\frac{1}{2}$–$10\frac{1}{2}$ lb);	cockerel	3.40–4.30 kg ($7\frac{1}{2}$–$9\frac{1}{2}$ lb)
Hen	3.20–4.10 kg (7–9 lb);	pullet	2.70–3.60 kg (6–8 lb)

Scale of points

Type and carriage	10
Indications of table merits	20
Indication of egg production merits	20
Colour and markings	20
Head	10
Legs and feet	10
Condition	10
	100

Defects

General coarseness. Superfine bone. Unfeathered shanks. Any points against table, laying or reproductive qualities. Birds may be passed for deformities, crooked breastbone, serious defects, and clean shanks devoid of any feathering.

BANTAM

North Holland Blue bantams should follow exactly the large fowl standard.

Weights

Male 1190 g (42 oz)
Female 1020 g (36 oz)

CARLISLE OLD ENGLISH GAME

LARGE FOWL

Origin: Great Britain
Classification: Hard feather
Egg colour: Tinted

When the Romans invaded Britain, Julius Caesar wrote in his commentaries that the Britons kept fowls for pleasure and diversion but not for table purposes. Many well-known authorities have considered that cock fighting was the diversion. In 1849 an Act of Parliament was passed making cock fighting illegal in this country, and with poultry exhibitions then taking root, many breeders began to exhibit Game fowls.

The Old English Game Club split in about 1930 as there was already a divergence of birds being shown with larger breasted, horizontally backed, exhibition type birds tending to win, and breeders of these formed the Carlisle Club, developing only some of the original colours. Breeders of the original type, wherein the back is at 45° to the ground, maintained the well-balanced, close-heeled, athletic fighting fowl, and formed the Oxford Club, retaining over thirty colours. The judge of Oxfords does so with the bird facing away from him to assess the correct balance. It is usually agreed that a good game fowl cannot be a bad colour.

General characteristics: male

Carriage: Bold, sprightly, the movement quick and graceful, as if ready for any emergency.
Type: Back short and flat, broad at the shoulders, and tapering to the tail. Breast broad, full, and prominent with large pectoral muscles, and breastbone not deep or pointed. Wings large, long and powerful with large strong quills, amply protecting the thighs. Belly small and tight. Tail large, carried upwards and spread, main feathers and quills large and strong.
Head: Small and tapering. Beak big, boxing (i.e. the upper mandible shutting tightly and closely over the lower one), crooked or hawk-like, pointed, strong at the base. Eyes large, bold and full of expression. Comb single, small, thin, upright, of fine texture in undubbed males and females. Face and throat skin flexible and loose. Ear-lobes and wattles fine, small and thin.
Neck: Large boned, round, of fair length, and very strong at the junction with the body, furnished with long and wiry feathers covering the shoulders.
Legs and feet: Legs strong. Thighs short, round and muscular, following the line of the body, or slightly curved. Shanks strong, clean boned, sinewy, close scaled (not flat and 'gummy'), not stiffly upright or too wide apart, and with a good bend or angle at the hock, fitted with hard and fine spurs set low. Toes, four, thin, straight and tapering, terminating in long, strong, curved nails, the fourth (or hind) toe strong, straight out and flat on the ground.
Plumage: Hard and glossy, without much fluff.
Handling: Well balanced, hard yet light fleshed, 'corky', with plenty of muscle, and

strong contraction of the wings and thighs to the body.

Female

With the exception of the tail, which is inclined to fan-shape and carried well up, the general characteristics are similar to those of the male, allowing for the natural sexual differences.

Colour

The black-red (partridge bred)

Male plumage: Neck, hackle and saddle dark rich red shading to deep orange. Breast and thighs black. Back and shoulders deep crimson. Wings, wing bow deep red with a rich dark blue bar across; secondaries bay colour on outer web; primaries black. Tail sound black with lustrous green gloss.

Female plumage: (Partridge) Neck golden-red and streaked with black. Breast and thighs shaded salmon. Back and wings partridge colour to be as free from rust and shaftiness as possible. Tail black with partridge coverts.

In both sexes: Legs white, yellow and willow. Face bright red: Eyes red, both alike.

The black-red (wheaten bred)

Male plumage: Neck, hackle bright orange, shading off to bright lemon. Back and saddle bright crimson. Breast and thighs black. Wings, wing bow bright red, in other respects, including tail, similar to (partridge) black-red.

Female plumage: (Wheaten) Neck, hackle clear red. Delicate, creamy self-colour on remainder except tail, which is nearly black. (Clay) Clay hens similar to wheaten only darker or harder in colour.

In both sexes: Legs white, occasionally yellow. Face bright red. Eyes red, both alike.

The brown-red

Male plumage: Neck and saddle lemon or orange, streaked with black. Breast black laced with brown to top of thighs. Back lemon or orange. Wings, shoulder and wing bow lemon or orange, rest of wing black. Tail black.

Female plumage: Neck, hackle lemon or orange, striped with black. Breast and thighs black laced with brown. Body black, tail black.

In both sexes: Legs dark. Face gypsy or red. Eyes dark, both alike.

The spangle

Male plumage: Neck, hackle and saddle dark red, finely tipped with white. Breast and thighs black, finely and evenly tipped with white. Back and shoulders dark red, finely tipped with white. Wings, wing bow dark red, finely tipped with white with a rich dark blue bar across, finely tipped with white; secondaries deep bay intermixed with white, bay predominating; primaries black intermixed with white. Tail, sickles and side hangers black, tipped with white, straight feathers black intermixed with white.

Female plumage: Neck, hackle golden-red, streaked with black, finely tipped with white. Breast and thighs dark salmon, finely and evenly tipped with white. Back and shoulders partridge coloured feathers, finely and evenly tipped with white. Wings, secondaries, partridge intermixed with white, partridge predominating; primaries dark, intermixed with white. All other feathers partridge colour finely and evenly tipped with white. Tail black with partridge coverts, finely and evenly tipped with white.

In both sexes: Legs white (occasionally yellow). Face bright red. Eyes red, both alike.

The birchen or grey

Male plumage: Neck, saddle grey or silver, streaked with black. Breast black laced with grey to top of thighs. Back grey or silver. Wings, shoulders and wing bow grey or silver, rest of wing black. Tail black.

Female plumage: Neck, hackle grey or silver, striped with black. Breast black laced with grey or silver to top of thigh. Body black, tail black.

In both sexes: Legs dark. Face gypsy or red. Eyes dark, both alike.

The golden duckwing

Male plumage: Hackle yellow, saddle straw. Breast and thighs black. Back and shoulders orange or rich yellow (golden). Wings, wing bow orange or rich yellow, wing bars steel-blue; secondaries white when closed, primaries black. Tail black.

Female plumage: Hackle silver white, finely striped with black. Breast pale salmon, diminishing to ash grey on thighs. Tail black, except top feathers which should match the body colour. Remainder French or steel grey, very lightly pencilled with black and even throughout.

In both sexes: Beak dark horn. Eyes ruby red. Comb, face, wattles and earlobes red. Legs and feet willow.

The silver duckwing

Male plumage: Neck and saddle white, free from dark streaks. Breast and thighs black. Back and shoulders silver-white. Wings, wing bow silver-white, wing bar steel-blue; secondaries white when closed; primaries black. Tail black.

Female plumage: As female golden duckwing.

In both sexes: Legs any self colour. Face bright red. Eyes red or pearl, both alike.

The blue-red

Male plumage: Neck, hackle and saddle orange or golden-red. Breast and thighs medium shade of blue. Back and shoulders deep or bright red. Wings, wing bow deep or bright red, with a rich dark blue bar across; secondaries bay colour on the outer web; primaries blue. Tail blue.

Female plumage: Neck golden-red, streaked with blue. Breast and thighs shaded salmon. Back and wings partridge colour, intermixed with blue. Tail to correspond with body colour.

In both sexes: Legs any self colour. Face bright red. Eyes red, both alike.

The blue-tailed wheaten hen

Male and female plumage: Similar in all respects to wheatens with the exception of wing primaries and tail shaded with blue. Legs white, occasionally yellow. Face bright red. Eyes red, both to be alike.

The crele

Male plumage: Neck and saddle chequered (barred) orange. Back and shoulders deep chequered orange. Wings, wing bow deep chequered orange with dark grey bar across; secondaries bay colour on the outer web; primaries dark grey. Tail dark grey.

Female plumage: Neck lemon chequered with grey. Breast and thighs chequered salmon. Back and wings chequered blue-grey. Tail to correspond with body colour.

In both sexes: Legs white preferred. Face bright red. Eyes red, both alike.

The cuckoo

Male plumage: Blue-grey barred, variations of this colour are: yellow, gold or red in the plumage.

Female plumage: Blue-grey barred all over.

In both sexes: Legs white preferred. Face bright red. Eyes red, both alike.

The pyle

Male plumage: Neck and saddle orange or chestnut-red. Breast and thighs white. Back and shoulders deep red. Wings, wing bow red with a white bar across; secondaries bay colour on outer web; primaries white. Tail white.

Female plumage: Neck lemon. Breast salmon, lighter towards thighs. Back and wings white. Tail white.

In both sexes: Legs white or yellow. Face bright red. Eyes red, both alike.

The self white

Male and female plumage: All over pure white. Legs white or yellow. Face bright red. Eyes red, both alike.

Duckwing Carlisle Old English Game male

Duckwing Carlisle Old English Game female

Brown-red Carlisle Old English Game male

Brown-red Carlisle Old English Game female

Grey Carlisle Old English Game male

Dark grey Oxford Old English Game male

Black-breasted red Oxford Old English Game, male

Crele Oxford Old English Game, male

Partridge Oxford Old English Game, female

Muffed partridge Oxford Old English Game, female

The self black

Male and female plumage: All over glossy black. Legs any self colour. Face red or dark. Eyes red or dark, both alike.

The self blue

Male and female plumage: Medium shade of blue. Legs any self colour. Face bright red. Eyes red, both alike.

Weights

Male	2.50–2.55 kg ($5\frac{1}{2}$–$5\frac{5}{8}$ lb). It is not considered desirable to breed males over 2.70 kg (6 lb)
Female	1.80–2.25 kg (4–5 lb)

Scale of points

Body (including breast, back and belly)	20
Handling (symmetry, cleverness, hardness of flesh and feathers, condition and constitution)	15
Head (including beak and eyes)	10
Neck	6
Shanks, spurs and feet	10
Plumage and colour	9
Thighs	8
Wings	7
Tail	6
Carriage, action and activity	9
	100

Serious defects

Thin thighs or neck. Flat sided. Deep keel. Pointed, crooked or indented breastbone. Thick insteps or toes. Duck feet. Straight or stork legs. In-knees. Soft flesh. Broken, soft or rotten plumage. Bad carriage or action. Any indications of weakness of constitution.

OXFORD OLD ENGLISH GAME

LARGE FOWL

Origin: Great Britain
Classification: Hard feather
Egg colour: Tinted

When the Romans invaded Britain, Julius Caesar wrote in his commentaries that the Britons kept fowls for pleasure and diversion but not for table purposes. Many well-known authorities have considered that cock fighting was the diversion. In 1849 an Act of Parliament was passed making cock fighting illegal in this country, and with poultry exhibitions then taking root, many breeders began to exhibit Game fowls. Over thirty colours of Old English Game have been known.

The Old English Game Club split in about 1930 as there was already a divergence of birds being shown with larger breasted, horizontally backed, exhibition type birds tending to win. Breeders of these formed the Carlisle Club, developing only some of the original colours. Breeders of the original type, wherein the back is at 45° to the ground, maintained the well-balanced, close heeled, athletic fighting fowl, and formed the Oxford Club, retaining over thirty colours. The judge of Oxfords does so with the bird facing away from him to assess the correct balance. It is usually agreed that a good game fowl cannot be a bad colour.

General characteristics: male

Carriage: Proud, defiant, sprightly, active on his feet, ready for any emergency, alert, agile, quick in his movements.
Type: Back short, flat, broad at the shoulders, tapering to the tail. Breast broad, full, prominent, with large pectoral muscles, breast bone not deep or pointed. Wings large, long and powerful with large strong quills, amply protecting the thighs. Tail large, up and spread, main feathers and quills large and strong. Belly small and tight. Thighs short round and muscular, following the line of the body, or slightly curved.
Head: Small and tapering, skin of face and throat flexible and loose. Beak big, boxing (i.e. the upper mandible shutting tightly and closely over the lower one), crooked or hawk-like, pointed, strong at the setting on. Eyes large, bold fiery and fearless. Comb, wattles and ear-lobes of fine texture, small and thin in undubbed males and females.
Neck: Large boned, round, strong, and of fair length, neck hackle covering the shoulders.
Legs and feet: Legs strong, clean boned, sinewy, close scaled, not fat and gummy like other fowls, close-heeled, not stiffly upright or too wide apart, and having a good bend or angle at the hock. Spurs are hard and fine and set low on the leg. Toes thin, long, straight and tapering, terminating in long, strong curved nails, hind toe of good length and strength, extending backwards in almost a straight line.
Plumage: Hard, sounding, resilient, smooth, glossy, and sufficient without much fluff.
Handling: Clever, well balanced, hard, yet light fleshed, corky, mellow and warm, with strong contraction of wings and thighs to the body.

Female

With the exception of the tail, which is inclined to fan-shape and carried well up, the general characteristics are similar to the male, allowing for natural sexual differences.

Colour

The black-breasted dark red

Male plumage: Hackle, shoulders and saddle rich dark red (the colour of the shoulders of a black-breasted red). The rest of the plumage black.
Female plumage: Body black or very dark brown, hackle ticked red.
In both sexes: Fluff (i.e. the down at the roots of the feathers next to the skin) black. Eyes, beak, legs and nails black. Face gypsy or purple.

The black-breasted red

Male plumage: Breast, thighs, belly and tail black. Wing bars steel-blue; secondaries (when closed) bay. Hackle and saddle feathers orange-red. Shoulders deep crimson-scarlet.
Female plumage: Hackle golden, lightly striped with black. Breast robin. Belly ash-grey. Back, shoulders and wings a good even partridge. Primaries and tail dark.
In both sexes: The dark legged birds should have grey fluff, the white and yellow legged, white fluff. Face scarlet red. Legs willow, yellow, white, carp or olive.

The shady or streaky-breasted light red

Male plumage: Hackle and back a shade lighter than the black breasted red male and sometimes red wing bars.
Female plumage: Wheaten, a pale cream colour (like wheat) with clear red hackle. Tail and primaries nearly black. The red wheaten (the colour of red wheat), or light brick-red in body and wings. Hackle dark red. Tail dark.
In both sexes: Fluff white. Legs white or yellow.

The black-breasted silver duckwing

Male plumage: Resembles the black-breasted red in his black markings and blue wing bars. Rest of the plumage clear, silvery white.

Female plumage: Hackle white, lightly striped black. Body and wings even silvery grey. Breast pale salmon. Primaries and tail nearly black.

In both sexes: Fluff light grey. Face red, eyes pearl. Legs and beak white. Or eyes red and legs dark.

The black-breasted yellow duckwing

Male plumage: Hackle and saddle yellow straw. Shoulders deep golden. Wing bars steel-blue; secondaries white when closed. Rest of plumage black.
Female plumage: Breast deeper, richer colour and body slightly browner tinge than the silver female.

In both sexes: Fluff light grey. Face red. Legs yellow, willow or dark.

The black-breasted birchen duckwing

Male plumage: Hackle deep rich straw, may be lightly striped. Shoulders maroon; otherwise same as preceding.
Female plumage: Shade darker than yellow duckwing female; hackle more heavily striped with black, and often foxy on the shoulders.

In both sexes: Face slightly darker than in yellow duckwing. Legs yellow or dark.

The black-breasted dark grey

Male plumage: Like the black-breasted red, except hackle, saddle and shoulders a dark, silvery grey, often striped with black.
Female plumage: Nearly black, with grey striped hackle, or body very dark grey.

In both sexes: Fluff black. Beak, eyes and legs black. Face gypsy or purple.

Other greys may have laced, streaked or mottled grey or throstle breasts. Hackle, saddle and shoulders more or less striped with black. Legs and eyes dark; the females dark grey to match. Fluff in both sexes, light or dark grey.
Note: Greys all differ from duckwings in having the secondaries, when closed, black, or, if grey, wanting the steel-blue bar across them.

The clear mealy-breasted mealy grey

Male plumage: Nearly white breasted, with hackle and saddle the same, lightly striped. Plumage and most of the tail grey.
Female plumage: Light grey.

In both sexes: Fluff light grey. Eyes and legs dark.

The brown-breasted brown-red

Male plumage: Breast, thighs, belly and closed wing mahogany-brown. Hackle and saddle almost similar. Shoulders crimson. Primaries and tail black or dark bronze-brown.
Female plumage: Dark mottled brown with light shafts to the feathers.

In both sexes: Fluff black. Face deep crimson or purple. Eyes and legs dark.

The streaky-breasted orange-red

Male plumage: Breast streaked, laced or pheasant, black, marked with brown or copper colour. Hackle and saddle brassy or coppery orange colour. Shoulders crimson, the rest of the wings and the tail black.
Female plumage: Black or nearly black body, with tinsel hackle striped with black, or dark mottled brown and gold striped hackle.

In both sexes: Fluff black or nearly so. Face, eyes and legs dark.

The ginger-breasted ginger-red

Male plumage: Breast and thighs deep yellow ochre, either clear or slightly pencilled or spotted. Hackle and saddle red-golden. Shoulders crimson-red. Tail and flight feathers bronzy.
Female plumage: Golden yellow throughout, pencilled or spangled, particularly on back and wings, with bronze. Tail pencilled bronze or dark.

In both sexes: Fluff dark. Beak, legs and eyes dark or yellow. Face purple or crimson.

The dun-breasted blue dun

Male plumage: Breast, belly, thighs, tail and closed secondaries the colour of a new slate, sometimes the breast marked with the same colour two shades darker. Hackle, saddle and shoulders, and sometimes the tail coverts and the primaries two shades darker (like a slate colour after being wetted).
Female plumage: Blue-slate colour, with dark hackle like the male, often marked or laced all over with the darker shade.
In both sexes: Fluff slate-blue. Eyes, face and legs dark.

The streaky-breasted red dun

Male plumage: Breast slate, streaked with copper-red. Hackle and saddles striped with slate or dark striped. Shoulders crimson. Wing bars and closed secondaries slate, or marked a little with brown. Tail slaty or dark blue.
Female plumage: Body slaty all over, or laced in a darker shade. Hackle golden striped, and sometimes marked with gold on the breast.
In both sexes: Fluff dark slate. Legs dark or yellow.

The yellow, silver and honey dun

These are coloured, respectively, with the following colours; the colour of new honeycomb is intended to describe the honey dun. They may have yellow or dark legs according to body colour, and white legs are permissible in the silver dun, as well as other coloured legs. The females are blue bodied with hackles to match their males. Smoky duns are of a dull smoke colour throughout; legs and eyes should be dark.

The pyle

Male plumage: The smock-breasted blood wing pyle is marked exactly like the black-breasted light red, except that the black and the blue wing bars are exchanged for a clear creamy white. The breast may be streaked with red in red pyle.
Female plumage: White, with salmon breast and golden striped hackle, or streaked all over lightly with red.
In both sexes: Face and eyes red. Legs white, yellow or willow.
Note: Other varieties of pyles may be streaky, marbled or robin breasted; and light lemon or custard in top colour, or dun pyles having slate-blue markings in place of red. All pyles have white fluff.

The spangled

Male and female plumage: These have white tips to their feathers. The more of these spots and the more regularly they are distributed the better. The male should show white ends to the feathers on hackle and saddle. The ground colour may be red, black or brown, or a mixture of all three. Underfluff white.
In both sexes: Eyes and face red. Legs any colour or mottled to match plumage.

The white

Male and female plumage: This variety should be free from any coloured feathers. Fluff pure white.
In both sexes: Beak and legs white. Face red, eyes pearl; or yellow legs and red eyes.

The black

Male and female plumage: This variety should be free from any white or coloured feathers and should possess black fluff.
In both sexes: Dark beaks, faces and legs and black eyes, though red faces and red eyes are allowed at present.

The furness, brassy back and polecat

Male and female plumage: These are blacks with brass colour on their wings or back, and occasionally have yellow legs, which are allowed. The females are chiefly black, but often much streaked with grey-brown on breast and wings. Polecats are streaked with dark tan colour on hackles and saddle in the males. Legs dark.

The cuckoo

Male and female plumage: Cuckoo-breasted cuckoo resembles the Plymouth Rock fowl in markings of a blue-grey barred plumage.

In both sexes: Faces and eyes red. Legs various.

Variations of this colour are yellow cuckoos, also creles, creoles, cirches, mackerels in different provincial dialects, having some mixture of gold or red in the plumage and white fluff, often extremely pretty. Legs white or yellow.

The brown-breasted yellow birchen

Male plumage: Breast reddish-brown. Hackles and saddle straw, striped birchen brown. Shoulders old gold or birchen. Wing bar and closed secondaries brown. Tail brown or bronze-black.

Female plumage: Yellow-brown, with grey hackle and robin breast.

In both sexes: Fluff light grey. Beak, legs and eyes yellow.

The hennie

Male and female: Hencocks should in their plumage resemble hens as closely as possible. They should have their hackle and saddle feathers rounded and the tail coverts hen-like, and not have much sheen on their feathers. This breed often runs large and reachy, which is one of its characteristics. The two centre tail feathers should be straight.

The muff and tassel

Male and female plumage: Both muffs and tassels, or topins, are recognized by the Oxford Club, there being famous strains of both, though now scarce. Tassels vary from a few long feathers (or lark tops) behind the comb to a good-sized bunch. They also occur in some strains of hennies. Muffs of the old breed are stronger, heavier boned birds than the males bred today, and are rather loose in feather.

Notes:

(1) It is desirable that the toenails should match the legs and beak in colour in all Game Fowl.
(2) White or yellow legged birds may have white feathers in wings and tail.
(3) The fancier, when he speaks of a brown-red, means the streaky-breasted orange-red; and when talking of a black-red, intends one to infer a black-breasted light red; while black-breasted dark greys are erroneously called 'birchens', although they have no birchen colour in them.

Weights

Male 1.80–2.50 kg (4–5½ lb) It is not considered desirable to breed males over 2.70 kg (6 lb)

Female 0.90–1.36 kg (2–3 lb)

Scale of points

Body (including breast, back and belly)	20
Handling (symmetry, cleverness, hardness of flesh and feathers, condition and constitution)	15
Head (including beak and eyes)	10
Neck	6
Shanks, spurs and feet	10
Plumage and colour	9
Thighs	8
Wings	7
Tail	6
Carriage, action and activity	9
	100

Serious defects

Thin thighs or neck. Flat sided. Deep keel. Pointed, crooked or indented breastbone. Thick insteps or toes. Duck feet. Straight or stork legs. In-knees. Soft flesh. Broken, soft or rotten plumage. Bad carriage or action. Any indication of weakness of constitution.

OLD ENGLISH GAME BANTAM

Classification: Hard feather

This standard is compiled from that of the O.E.G. Bantam Club, and follows the Oxford ideal. Other standards exist, but essential differences are slight. Chief variations are in methods of interpretation. O.E.G. bantams are of comparatively recent creation. They were evolved largely from the common crossbred bantam of the countryside. Probably there is very little large breed blood in them.

In the large breed it is usually agreed that a good Game bird cannot be a bad colour. This remark does not apply bantams, which are show birds only, colour playing a very important part. Nevertheless, the ideal is that the bantam should be a true miniature of the national fighting Game – though this is seldom the case.

General characteristics: male

Carriage: Bold, sprightly, defiant and proud, active on feet, agile, quick in movement, ready for any emergency.
Type: Body short. Black flat, broad at shoulders, tapering to tail. Breast broad, full and prominent, with strong pectoral muscles and breastbone not deep or pointed. Belly small and tight. Wings strong and powerful and carried well up. Tail large, carried upwards and spread, quills and main feathers large and strong.
Head: Small and tapering, skin of face and throat flexible and loose. Beak big and boxing (the upper mandible shutting tightly over the lower), crooked or hawk-like, pointed, strong at base. Eyes large, bold, fiery and alike in colour. Comb single, small, upright and thin. Ear-lobes and wattles fine, small and thin.
Neck: Large boned, round, strong and fairly long, neck hackle long and wiry and covering the shoulders.
Legs and feet: Legs strong, clean boned, sinewy, close scaled, not fat and gummy, not stiffly upright, not too wide apart, with a good angle at the hock. Thighs short, round and muscular, following line of body, slightly curved. Toes, four, long, straight and tapering: hind toe of good length and strength, straight and firm on ground. Spurs hard and fine, set low.
Plumage: Hard, resilient, smooth and glossy, without much fluff.
Handling: Well balanced, mellow, firm and corky, hard yet light fleshed, with strong contraction of wings and thighs to body.

Female

The general characteristics are similar to those of the male, allowing for the natural sexual differences. The tail is inclined to fan-shape and carried well up.

Colour

Colours are very numerous, the most popular being spangles, black-reds (wheaten bred and partridge bred), duckwings, brown-reds, self blacks and blues, furnesses, creles and greys. Most of these main colours follow the usual colour-pattern applicable to the variety.

Furnesses are black with brassy hackles, wing shoulders and backs in males; and black with greyish-brown streaked breasts and wings in females. Creles are barred varieties of other colours, distinct from cuckoos, which are plain barred grey. In black-reds, the wheaten bred is brighter in colour and more popular than partridge bred. Off colours are very numerous, but not so popular as in the large breed. Pyles and whites are seldom seen. Hennies (hen-feathered males), muffs and tassels (beards and top-knots) are recognized sub-varieties in all colours.

Colour of face, beak, eyes, shanks and toenails varies with the particular colour-variety: from red face, red eyes, white legs, toenails and beak in spangles to deep crimson, purple or black faces, black eyes, black legs and feet in brown-reds and similar dark colours. Willow, olive and yellow legs are frequent and permissible in various sub-varieties.

Some of the more popular colour varieties, with the names usually applied to them by showmen, are given below.

The spangle

Male plumage: Neck, hackle and saddle dark red, finely tipped with white. Breast and thighs black, finely and evenly tipped with white. Back and shoulders dark red, finely tipped with white. Wings, wing bow dark red, finely tipped with white with a rich dark blue bar across, finely tipped with white; secondaries deep bay intermixed with white, bay predominating; primaries black intermixed with white. Tail, sickles and side hangers tipped with white, straight feathers black intermixed with white.

Female plumage: Neck, hackle golden-red, streaked with black, finely tipped with white. Breast and thighs dark salmon, finely and evenly tipped with white. Back and shoulders partridge coloured feathers, finely and evenly tipped with white. Wings, secondaries, partridge intermixed with white, partridge predominating; primaries dark, intermixed with white. All other feathers partridge colour finely and evenly tipped with white.

In both sexes: Legs white (occasionally yellow). Face bright red. Eyes red, both alike.

The black-red (partridge bred)

Male plumage: Neck, hackle and saddle dark rich red shading to deep orange. Breast and thighs black. Back and shoulders deep crimson. Wings, wing bow deep red with a rich dark blue bar across; secondaries bay colour on outer web; primaries black. Tail, sound black with lustrous green gloss.

Female plumage: (Partridge) Neck golden-red and streaked with black. Breast and thighs shaded salmon. Back and wings partridge colour to be as free from rust and shaftiness as possible. Tail, black with partridge coverts.

In both sexes: Legs white, yellow and willow. Face bright red. Eyes red, both alike.

The black-red (wheaten bred)

Male plumage: Neck, hackle bright orange, shading off to bright lemon. Back and saddle bright crimson. Breast and thighs black. Wings, wing bow bright red, in other respects, including tail, similar to (partridge) black-red.

Female plumage: (Wheaten) Neck, hackle clear red. Delicate creamy self colour on remainder except tail, which is nearly black. (Clay) Clay hens similar to wheaten only darker or harder in colour.

In both sexes: Legs white, occasionally yellow. Face bright red. Eyes red, both alike.

The golden duckwing

Male plumage: Hackle yellow, saddle straw. Breast and thighs black. Back and shoulders orange or rich yellow (golden). Wings, wing bow orange or rich yellow, wing bars steel-blue; secondaries white when closed, primaries black. Tail black.

Female plumage: Neck and saddle white, free from dark streaks. Breast and thighs black. Back and shoulders silver-white. Wings, wing bow silver-white, wing bar steel-blue; secondaries white when closed; primaries black. Tail black.

In both sexes: Legs any self colour. Face bright red. Eyes red or pearl, both alike.

Silver Duckwing Old English Game male, bantam

Furness Old English Game female, bantam

Black-red Old English Game male, bantam

Blue Old English Game female, bantam

Pile Old English Game female, bantam

The silver duckwing

Male plumage: Neck and saddle white, free from dark streaks. Breast and thighs black. Back and shoulders silver-white. Wings, wing bow silver-white, wing bar steel-blue; secondaries white when closed; primaries black. Tail black.

Female plumage: As female golden duckwing.

In both sexes: Legs any self colour. Face bright red. Eyes red or pearl, both alike.

The blue-red

Male plumage: Neck, hackle and saddle orange or golden-red. Breast and thighs medium shade of blue. Back and shoulders deep or bright red. Wings, wing bow deep or bright red, with a rich dark blue bar across; secondaries bay colour on the outer web; primaries blue. Tail blue.

Female plumage: Neck golden-red, streaked with blue. Breast and thighs shaded salmon. Back and wings partridge colour, intermixed with blue. Tail to correspond with body colour.

In both sexes: Legs any self colour. Face bright red. Eyes red, both alike.

The blue-tailed wheaten hen

Plumage: Similar in all respects to wheatens with the exception of wing primaries and tail shaded with blue. Legs white, occasionally yellow. Face bright red. Eyes red, both to be alike.

The brown-red

Male plumage: Neck and saddle lemon or orange, streaked with black. Breast black laced with brown to top of thighs. Back lemon or orange. Wings, shoulder and wing bow lemon or orange, rest of wing black. Tail black.

Female plumage: Neck, hackle lemon or orange, striped with black. Breast and thighs black laced with brown. Body black, tail black.

In both sexes: Legs dark. Face gypsy or red. Eyes dark, both to be alike.

The birchen or grey

Male plumage: Neck, saddle grey or silver, streaked with black. Breast black laced with grey to top of thighs. Back grey or silver. Wings, shoulders and wing bow grey or silver, rest of wing black. Tail black.

Female plumage: Neck, hackle grey or silver, striped with black. Breast black laced with grey or silver to top of thigh. Body black, tail black.

In both sexes: Legs dark. Face gypsy or red. Eyes dark, both to be alike.

The crele

Male plumage: Neck and saddle chequered (barred) orange. Back and shoulders deep chequered orange. Wings, wing bow deep chequered orange with dark grey bar across; secondaries bay colour on the outer web: primaries dark grey. Tail dark grey.

Female plumage: Neck lemon chequered with grey. Breast and thighs chequered salmon. Back and wings chequered blue-grey. Tail to correspond with body colour.

In both sexes: Legs white preferred. Face bright red. Eyes red, both to be alike.

The cuckoo

Male plumage: Blue-grey barred, variations of this colour are: yellow, gold or red in the plumage.

Female plumage: Blue-grey barred all over.

In both sexes: Legs white preferred. Face bright red. Eyes red, both to be alike.

The pyle

Male plumage: Neck and saddle orange or chestnut-red. Breast and thighs white. Back and shoulders deep red. Wings, wing bow red with a white bar across; secondaries bay colour on outer web; primaries white. Tail white.

Female plumage: Neck lemon. Breast salmon, lighter towards thighs. Back and wings white. Tail white.

In both sexes: Legs white or yellow. Face bright red. Eyes red, both to be alike.

The furness

Male plumage: Black, with brassy wing butts, wing bays, saddle hangers, brassiness to be allowed in hackles and on back.

Female plumage: Black with salmon breast, brassiness allowed on wings, back, cushion and hackles.

In both sexes: Leg colour any self colour, both alike. Eyes red. Face bright red.

The blue furness

Male plumage: Blue, with brassy coloured wing butts, wing bays, saddle hangers, brassiness allowed in neck hackles and on back.

Female plumage: Blue, with salmon breast, brassiness allowed on wings and back, hangers and hackles.

In both sexes: Leg colour any self colour, both alike. Face and eyes red.

The brassy-backed black

Male plumage: Black, with brass colour on back, wing butts, hackles and hangers. No bay colour allowed. Clear hackles and hangers allowed.

Female plumage: Polecat. Pale brown to dark brown with black neck hackle, tail and wing flights, brown or brassy streaks allowed in neck hackle. Some speckles and/or lacing allowed on breast and belly.

In both sexes: Legs to be any self colour, both alike. Face and eyes red.

The brassy-backed blue

Male plumage: Blue with brass colour on back, wing butt, hackles and hangers. No wing bay colour. Clear hackles and hangers allowed.

Female plumage: Dun. Blue with light to dark brown or honey, or brassiness in neck hackle, cushion, back, wings, breast and belly. Speckling and lacing allowed on breast and belly.

In both sexes: Legs any self colour, both alike. Face and eyes red.

The blue duckwing

Male plumage: Cream neck and saddle free from dark streaks. Breast and thighs blue. Back and shoulder silver-white. Wings, wing bay silver-white, wing bar steel-blue; secondaries white when closed; primaries blue. Tail blue.

Female plumage: Neck silver lights striped with black. Breast salmon shading off to blue on thighs. Back and wings silver-grey free from rust and shaftiness.

In both sexes: Legs any self colour, both alike. Face red. Eyes red or pearl.

The lemon blue

Male plumage: Neck and saddle lemon or orange streaked with blue, breast blue, laced with lemon to top of thighs. Back lemon to orange. Wings, shoulders and wing bay lemon or orange, rest of wings blue. Tail blue.

Female plumage: Neck hackle lemon or orange striped with blue. Breast and thighs blue, laced with brown. Body blue, tail blue.

In both sexes: Legs dark. Face gypsy or red. Eyes red, both to be alike.

The splashed

Male and female plumage: Mixture of black, white and blue, giving an overall appearance of being light grey.

In both sexes: Legs any self colour. Face red. Eyes red, both to be alike.

The blue grey

Male and female plumage: Silver where the lemon blue are lemon. Legs any self colour, both alike. Face red. Eyes red.

The ginger-red

Male plumage: Neck and saddle red-golden. Breast and thighs, deep yellow, clear or slightly pencilled or spotted. Back and shoulders crimson-red. Wings and tail bronze.

Female plumage: Golden-yellow throughout; pencilled or spangled particularly on back and wings with bronze. Tail pencilled bronze or dark, fluff dark.

In both sexes: Legs dark or yellow. Face dark or red. Eyes dark or red, both to be alike.

The self white

Male and female plumage: All over pure white. Legs white or yellow. Face bright red. Eyes red, both to be alike.

The self black

Male and female plumage: All over glossy black. Legs any self colour. Face red or dark. Eyes red or dark, both to be alike.

The self blue

Male and female plumage: Medium shade of blue. Legs any self colour. Face bright red. Eyes red, both alike.

Muffs and tassels

Both muffs and tassels are recognized. Tassels vary from a few long feathers behind the comb to a good-sized bunch.

Weights (suggested)

Male	620–740 g (22–26 oz)
Female	510–620 g (18–22 oz)

Scale of points

Head, beak and eyes	10
Neck	6
Body, breast, back and belly	20
Wings	7
Tail	6
Thighs	8
Shanks, spurs and feet	10
Plumage and colour	9
Handling (symmetry, cleverness, hardness of flesh and feathers, condition and constitution)	15
Carriage, action and activity	9
	100

Serious defects

Thin thighs or neck. Flat sided; deep keel. Pointed, crooked or indented breastbone. Thick insteps or toes. Duck feet. Straight or stork legs, or in-kneed. Soft flesh. Broken, soft or rotten plumage. Bad carriage or action. Any indication of weakness of constitution.

OLD ENGLISH PHEASANT FOWL

LARGE FOWL

Origin: Great Britain
Classification: Light: Rare
Egg colour: White

This breed was given its name of Old English Pheasant Fowl about 1914, previous to which it had been called the Yorkshire Pheasant, Golden Pheasant and also the Old-fashioned Pheasant. That it is a very old English breed is certain. Some Northern breeders retained their strains as Yorkshire Pheasant Fowls until the present tag of 'Old English' was brought officially into use. It has a meaty breast for a light breed, and has always been popular with farmers.

General characteristics: male

Carriage: Alert and active.
Type: Body rather long, deep and round with prominent shoulders. Tail flowing and set well back.
Head: Fine. Beak of medium size. Eyes bright and prominent. Comb rose, moderate in size, not impeding either sight or breathing, fine texture, evenly set, rather square front, the top flat and with plenty of work, tapering to a single leader (or spike) at the back, which should curve gracefully downwards, following the neck line but quite free from it. Face and wattles smooth, free from coarseness or wrinkles. Ear-lobes medium size, oval or almond shape, smooth.
Neck: Graceful.
Legs and feet: Legs of medium length, well apart, neither coarse nor too fine. Shanks free from feathers. Toes, four, well spread.

Female

With the exception of the tail (moderately whipped) the general characteristics are similar to those of the male, allowing for the natural sexual differences.

Old English Pheasant Fowl large fowl Male

Colour

The gold

Male plumage: Ground colour bright rich bay. Back rich mahogany-red. Lacing, bars, striping, tipping and tail beetle green-black. Hackles striped and slightly tipped. Saddle a slightly deeper shade than neck. Breast laced. Wing bars (two) marked.

Female plumage: Ground colour bright rich bay. Striping, tipping, spangling and tail beetle green-black. Neck with heavy stripe down centre of each feather. Wing bars crescent spangling; even and well-marked bars a point of great beauty. Tail with slight edging of ground colour carried up from the base along the upper edge of the tail. Remainder, each feather tipped with a crescent-shaped spangle. Shafts of all feathers bay.

The silver

Male and female plumage: White with beetle green-black markings.

In both sexes and colours

Beak horn. Eyes fiery red. Comb bright rich red. Face and wattles red. Ear-lobes white. Legs and feet slate-blue.

Old English Pheasant Fowl large fowl Female

Weights

Cock	2.70–3.20 kg (6–7 lb);	cockerel	2.50–2.70 kg ($5\frac{1}{2}$–6 lb)
Hen	2.25–2.70 kg (5–6 lb);	pullet	2.00–2.25 kg ($4\frac{1}{2}$–5 lb)

Scale of points

Type (including legs)	20
Head (comb 15, lobes 5, other points 5)	25
Markings	20
Ground colour of body	15
Plumage and flow of feather	10
Size	5
Condition	5
	100

Serious defects

Comb single or over either side. Blushed lobes. Sooty hackles. Definitely black breast in male. Superfine bone. Lack of size. Squirrel tail. Any deformity. Any other defects which would affect health, hardiness, productivity or activity up to 20 points.

BANTAM

Old English Pheasant Fowl bantams should follow exactly the large fowl standard.

Weights

Male	790 g (28 oz)
Female	680 g (24 oz)

ORLOFF

LARGE FOWL

Origin: Iran and Russia
Classification: Heavy: Rare
Egg colour: Tinted

This breed originally came from the Gilan province of northern Iran, where it was known as the Chilianskaia. Some were taken to Moscow and renamed by Count Orloff Tech-esmensky. From Russia they became known to British, Dutch and German poultry experts in the 1880s and 1890s, and an Orloff Club existed in Britain in the 1920s and 1930s. Orloff bantams have been known in Germany since 1925, but did not reach Britain until the 1970s. As this standard indicates, Orloffs are mainly judged on type and character, especially of the head. Plumage colour is a secondary consideration.

General characteristics: male

Carriage: Upright with slightly sloping back.
Type: Body broad and fairly long. Flat, slightly sloping back. Breast rather full and prominent. Closely carried wings of moderate length. Tail of medium size with fairly narrow sickles. Carriage rather low but slightly above horizontal.
Head: Skull wide, of medium size. Beak short, stout and well hooked. Eyes full, and deeply set under well-projecting (beetle) eyebrows, giving a gloomy, vindictive expression. Comb low and flat, shaped somewhat like a raspberry cut through its axis (lengthwise), covered with small protuberances mingled with small, bristle-like feathers, which peculiarity is particularly noticeable in the female. Face muffled, beard and whiskers well developed. Ear-lobes very small, hidden under the muffles. Wattles small, and show only in the male.
Neck: Fairly long and erect, very heavily covered with hackle (boule), the feathers very full at the top but so close at the base of the neck as to appear thin there, and forming a distinct angle with the back.
Legs and feet: Moderately long and stout. Thighs muscular and well apart. Shanks round and finely scaled. Toes, four, long and well spread.

Female

With the exception of the muffling (which is more developed) and the tail (comparatively long) the general characteristics are similar to those of the male, allowing for the natural sexual differences.

Colour

The black

Male and female plumage: Solid black to the skin from head to tail, with a beetle-green sheen.

The cuckoo

Male and female plumage: As in Cuckoo Leghorn.

The mahogany

Male plumage: Beard and whiskers a mixture of black, mahogany and grey, grey preponderating. Neck hackle rich dark orange to mahogany, darkest at the crown and showing very slight black stripes at the base only. Saddle rich mahogany shading to deep orange. Wings rich deep mahogany with a strongly defined green-black sheen.
Female plumage: Mufflings as in the male. Hackle mahogany, the lower feathers showing black striping. Tail mainly black. Remainder rich dark mahogany uniformly peppered with black, the entire absence of black, or heavy and irregular black splashes undesirable.

Spangled Orloff male, large

The spangled

Male plumage: Hackles rich orange, with white tips to as many feathers as possible. Back rich mahogany. Wings rich mahogany with black bar showing green or purple sheen, and white flights. Breast solid black with white tips, blotchiness or washiness undesirable. Tail green-black.
Female plumage: Light mahogany with white tips, the spangling to be as uniform as possible.

The white

Male and female plumage: Lustrous white from head to tail.

In both sexes and all colours

Beak yellow, with a thin rose-tinted skin at base of beak and nostrils. Eyes red or amber. Comb, face, ear-lobes and wattles red. Legs rich yellow.

Weights

Cock	3.60 kg (8 lb);	cockerel	3.20 kg (7 lb)
Hen	2.70 kg (6 lb);	pullet	2.25 kg (5 lb)

Scale of points

Type and carriage	25
Colour	20
Comb and other head points including muffling	35
Legs	10
Condition	10
	100

Serious defects

Absence of beard and muffling, and puffed hackle. Legs other than yellow. Comb of any other form than as described. Weak, deformed or diseased specimens.

Disqualifications

In this breed the colour is of secondary importance and is a deciding point only in close competition. The main characteristics of the Orloff are its peculiarities of shape, comb, head and carriage and judges are earnestly requested to bear this in mind when awarding prizes. Slight feathering or down between the toes is not to constitute a disqualification.

BANTAM

Orloff bantams follow the large fowl standard.

Weights

Cock	1175 g (40 oz);	cockerel	1000 g (36 oz)
Hen	1000 g (36 oz);	pullet	900 g (32 oz)

ORPINGTON

LARGE FOWL

Origin: Great Britain
Classification: Heavy: Soft feather
Egg colour: Brown

In the Orpington we have an English breed named after the village in Kent where the originator, William Cook, had his farm. He introduced the black variety in 1886, the white in 1889 and the buff in 1894. Within five years of the original black Orpington being introduced exhibition breeders were crossing Langshan and Cochin and exhibiting the offspring as black Orpingtons, the birds fetching high prices, and attracting many for their immense size. But this crossing at once turned a dual-purpose breed into one solely for show purposes, and it has remained so until today. A late introduction, the Jubilee Orpington, is now rarely seen.

General characteristics: male

Carriage: Bold, upright and graceful; that of an active fowl.
Type: Body deep, broad and cobby. Back nicely curved with a somewhat short, concave outline. Saddle wide and slightly rising, with full hackle. Breast broad, deep and well rounded, not flat. Wings small, nicely formed and carried closely to the body, the ends almost hidden by the saddle hackle. Tail rather short, compact, flowing and high, but by no means a squirrel tail.
Head: Small and neat, fairly full over the eyes. Beak strong and nicely curved. Eyes large and bold. Comb single, small, firmly set on head, evenly serrated and free from side sprigs. In the black variety, comb may be single or rose, the latter small, straight and firm, full of fine work or small spikes, level on top (not hollow in centre), narrowing behind to a distinct peak lying well down to the head (not sticking up). Face smooth. Wattles of medium length, rather oblong and nicely rounded at the bottom. Ear-lobes small and elongated.
Neck: Of medium length, curved, compact and with full hackle.
Legs and feet: Legs short and strong, the thighs almost hidden by the body feathers, well set apart. Toes, four, straight and well spread.

Plumage: Fairly profuse but close, not soft, loose and fluffy as in the Cochin, or close and hard as in the Game Fowl.
Handling: Firm.

Female

The general characteristics are similar to those of the male. Her cushion should be wide but almost flat, and slightly rising to the tail, sufficient to give the back a graceful appearance with an outline approaching concave.

Colour

The blue

Male plumage: Hackles, saddle, wing bow, back and tail dark slate-blue. Remainder slate-blue, each feather to show lacing of darker shade as on back.
Female plumage: Medium slate-blue, laced with darker shade all through, except head and neck, dark slate-blue.

In both sexes: Beak black. Eyes black or very dark brown, black preferred. Comb, face, wattles and ear-lobes bright red. Legs and feet black or blue. Toenails white.

The black

Male and female plumage: Black with a green sheen.

In both sexes: Beak, etc. as in the blue. Soles of feet white.

The buff

Male and female plumage: Clear, even buff throughout to the skin.

In both sexes: Beak white or horn. Eyes red or orange colour. Comb, face, ear-lobes and wattles bright red. Legs, feet and toenails white. Skin white.

The white

Male and female plumage: Pure snow-white.

In both sexes: Beak, legs, feet and skin white. Eyes, face, ear-lobes and wattles red.

Weights

The blue

Male	Not less than 4.55 kg (10 lb)
Female	Not less than 3.40 kg (7½ lb)

The black

Male	Not less than 4.55 kg (10 lb)
Female	Not less than 3.60 kg (8 lb)

The buff and white

Male	Not less than 3.60 kg (8 lb)
Female	Not less than 2.70 kg (6 lb)

Scale of points

The blue

Type	25
Size (with utility qualities)	20
Head	10
Legs and feet	10
Colour and plumage	25
Condition	10
	100

Buff Orpington male, large

Buff Orpington female, large

Black Orpington male, large

Blue Orpington female, large

Black Orpington male, bantam

The black

Type (shape) (body 15, breast 10, saddle 5)	30
Size	10
Carriage	10
Head (skull 5, comb 7, face 5, eyes 5, beak 3)	25
Skin	5
Legs and feet	5
Plumage and condition	10
Tail	5
	100

The buff and white

Type	30
Size	10
Head	15
Legs and feet	10
Colour	20
Condition	15
	100

Serious defects

Side sprigs on comb. White in ear-lobes. Feather or fluff on shanks or feet. Long legs. Any deformity. Yellow skin or yellow on the shanks or feet of any variety. Any yellow or sappiness in the white. Coarseness in head, legs or feathers of the buff.

Disqualifications

Trimming or faking.

BANTAM

Orpington bantams are miniatures of their large fowl counterparts and the standard for those should be followed, but a rose comb is not standardized in bantam.

Weights

Male 1.70 kg ($3\frac{3}{4}$ lb)
Female 1.50 kg ($3\frac{1}{4}$ lb)

Scale of points

	Black	*Blue*	*Buff and white*
Type, carriage and feather	35	30	30
Colour and undercolour	15	30	20
Head and eyes	20	15	15
Legs, feet and skin	15	10	10
Size and condition	15	15	25
	100	100	100

PEKIN BANTAM

Origin: Asia
Classification: True bantam
Egg colour: White or cream

This is a genuine bantam breed, very old and having no real relationship to the large breed of Cochins. It was imported from Pekin in the middle of the nineteenth century, hence its name. In recent years new colours have been added to the standard.

General characteristics: male

Carriage: Bold, rather forward and low, the head very little higher than the tail.
Type: Body short and broad. Back short, increasing in breadth to the saddle, which should be very full, rising well from between the shoulders and furnished with long soft feathers. Breast deep and full. Wings short, tightly tucked up, the ends hidden by saddle hackle. Tail very short and full, soft and without hard quill feathers, with abundant coverts almost hiding main tail feathers, the whole forming one unbroken duplex curve with the back and saddle. General type: tail should be carried higher than the head – 'tilt'.
Head: Skull small and fine. Beak rather short, stout, slightly curved. Eyes large and bright. Comb single, small, firm, perfectly straight and erect, well serrated, curved from front to back. Face smooth and fine, ear-lobes smooth and fine, preferably nearly as long as the wattles, which are long, ample, smooth and rounded.
Neck: Short, carried forward, with abundant long hackle reaching well down the back.
Legs and feet: Legs short and well apart. Stout thighs hidden by plentiful fluff. Hocks completely covered with soft feathers curling round the joints (stiff feathers forming 'vulture hocks' are objectionable but not a disqualification). Shanks short and thick, abundantly covered with soft outstanding feathers. Toes, four, strong and straight, the middle and outer toes plentifully covered with soft feathers to their tips.
Plumage: Very abundant, long and wide, quite soft with very full fluff.

Female

With the exception of the back (rising into a very full and round cushion) the general characteristics are similar to those of the male, allowing for the natural sexual differences.

Cuckoo Pekin male

Columbian Pekin female

White Pekin male

Lavender Pekin female

Colour

The black

Male and female plumage: Rich sound black with lustrous beetle-green sheen throughout, free of white or coloured feathers. (*Note:* Some light undercolour in adult males is permissible as long as it does not show through.)

The blue

Male and female plumage: A rich pale blue (pigeon blue preferred) free from lacing, but with rich dark blue hackles, back and tail in the male.

The buff

Male and female plumage: Sound buff, of a perfectly even shade throughout, quite sound to roots of feathers, and free from black, white or bronze feathers. The exact shade of buff is not material so long as it is level throughout and free from shaftiness, mealiness or lacing. (*Note:* A pale 'lemon buff' is usually preferred in the show-pen.)

The cuckoo

Male and female plumage: Evenly banded with dark slate on light French grey ground colour.

The mottled

Male and female plumage: Evenly mottled with white at the tip of each feather on a rich black with beetle-green sheen.

The barred

Male and female plumage: Each feather barred across with black bars, having a beetle-green sheen on a white background. The barring to be equal proportions of black and white. The colours to be sharply defined and not blurred or shaded off. Barring should continue through the shaft and into the underfluff, and each feather must finish with a black tip. Plumage should present a bluish, steely appearance free from brassiness and of a uniform shade throughout.

In both sexes: Eyes red orange. Legs and feet yellow.

The columbian

Male and female plumage: Pearl-white with black markings. Head and neck white with dense black stripe down middle of each feather, free from black edgings or black tips. Saddle pearl-white. Tail feathers and tail coverts glossy green-black, the coverts laced or not with white. Primaries black, or black edged with white; secondaries black on inner edge, white outer. Remainder of plumage entirely white, of pearl-grey shade, free from ticking. Undercolour either slate, blue-white or white.

The lavender

Male and female plumage: The lavender is not a lighter shade of the blue Pekin. It is different genetically and is of a lighter more silver tint without the darker shade associated with the normal blue. The silver tint is most obvious in the neck and saddle hackle feathers of the male.

In both sexes: Beak yellow or horn. Eyes orange, red or yellow. Legs and toes deep yellow.

The partridge

Male plumage: Head dark orange-red, neck hackle bright orange or golden-red, becoming lighter towards the shoulders and preferably shading off as near lemon colour as possible, each feather distinctly striped down the middle with black, and free from shaftiness, black tipping or black fringe. Saddle hackle to resemble neck hackle as nearly as possible. Breast, thighs, underparts, tail, coverts, wing butts and foot feather, hock feather and fluff lustrous green-black, free from grey, rust or white. Back, shoulder coverts and wing bow rich crimson. Primaries black, free from white or grizzle; secondaries black inner web, bay outer, showing a distinct wing bay when closed.

Female plumage: Head and neck hackle light gold or straw, each feather distinctly striped down middle with black. Remainder clear light partridge brown, finely and evenly pencilled all over with concentric rings of dark shade (preferably glossy green-black). The whole of uniform shade and marking, and the ground colour of the soft brown shade frequently described as the colour of a dead oak leaf, with three concentric rings of pencilling or more over as much of the plumage as possible.

The white

Male and female plumage: Pure snow-white, free from cream or yellow tinge, or black splashes or peppering.

In both sexes and all colours

Beak yellow, but in dark colours may be shaded with black or horn. Eyes red, orange or yellow – red preferred. Comb, face, wattles and ear-lobes bright red. Legs and feet yellow. (Dark legs permissible in blacks if the soles of the feet and back of shanks are yellow.)

Weights

Male	680 g (24 oz) max.
Female	570 g (20 oz) max.

Scale of points

Colour and markings	15
Fluff and cushion	15
Leg and foot feather	10
Size and weight	10
Type and carriage	20
Head	10
Length of shank	10
Condition	10
	100

Serious defects

Twisted or drooping comb. Slipped wings. Legs other than yellow (except for blacks). Eyes other than red, orange or yellow. Any deformity. Split front undesirable, but not a defect.

PLYMOUTH ROCK

LARGE FOWL

Origin: America
Classification: Heavy: Soft feather
Egg colour: Tinted

Specimens of the barred Plymouth Rock were first exhibited in America in 1869, and stock reached here in 1871. The white and black varieties came as sports. About 1890, the buff was exhibited in America and in England. The barred Rock came to us as a dual-purpose breed, but was developed to an exhibition ideal in which body size and frontal development were neglected in order to secure long narrow finely barred feathers. With the introduction of sex-linkage between the black Leghorn and barred Rock for commercial purposes, utility breeders made use of the Canadian barred Rock, a bird with roomy body, full breast, lower on the leg but coarser in barring.

Barred Plymouth Rock male, large

Barred Plymouth Rock female, bantam

White Plymouth Rock male, bantam

Buff Plymouth Rock female, bantam

General characteristics: male

Carriage: Alert, upright with bold appearance, well balanced and free from stiltiness.
Type: Body large, deep and compact, evenly balanced and symmetrical, broad, the keel bone long and straight. Back broad and of medium length, saddle hackle of good length and abundant. Breast broad and well rounded. Wings of medium size, carried well up, bow and tip covered by breast and saddle feathers, respectively; flights carried horizontally. Tail medium size, rising slightly from the saddle to be carried neatly and not to be fan, squirrel, or wry tail, sickles medium length and nicely curved, coverts sufficiently abundant to cover the stiff feathers.
Head: Of medium size, strong and carried well up. Beak short, stout and slightly curved. Eyes large, bright and prominent. Comb single, medium in size, straight and erect with well-defined serrations, smooth and of fine texture, free from side sprigs and thumb marks. Face smooth. Ear-lobes well developed, pendant, and of fine texture. Wattles moderately rounded and of equal length, to correspond with size of comb, smooth and of fine texture.
Neck: Of medium length, slightly curved, a full hackle flowing over the shoulders.
Legs and feet: Legs wide apart. Thighs large and of medium length. Shanks medium length, stout, well rounded, smooth and free from feathers. Toes, four, strong and perfectly straight, well spread and of medium length.
Fluff: Moderately full, carried closely to the body and of good texture.
Skin: Silky and fine in texture.

Female

The general characteristics are similar to those of the male, allowing for the natural sexual differences, except that comb, ear-lobes and wattles are smaller, the neck is of medium length, carried slightly forward, and the tail is small and compact, carried well back.

Colour

The barred

Male and female plumage: Ground colour, white with bluish tinge, barred with black of a beetle-green sheen, the bars to be straight, moderately narrow, of equal breadth and sharply defined, to continue through the shafts of the feathers. Every feather to finish with a black tip. The fluff, or undercolour, to be also barred. The neck and saddle hackles, wing bow and tail to correspond with the rest of the body, presenting a uniformity of colour throughout.

In both sexes: Eyes rich bay. Legs and beak yellow. Face, ear-lobes, wattles and comb red.

The black

Male and female plumage: Black with a beetle-green sheen. Eyes etc. as in the barred.

The buff

Male and female plumage: Clear, sound, even golden-buff throughout to the skin. The tail clear buff to harmonize with the body colour, and the undercolour (or fluff) and the quill of the feather also to harmonize with the surface colour. The male of more brilliant lustre than the female. Eyes etc. as in the barred.

The columbian

Male plumage: Head pure white. Neck hackle white with a distinct broad black stripe down the centre of each feather, free from white shaft, such stripe to be entirely surrounded by a clearly defined white margin, finishing with a decided white tip, free from black outer edging, black tips and excess of greyness at throat. Saddle hackle pure white. Tail, main feathers black with beetle-green sheen; coverts black with beetle-green sheen either laced or not with white. Wings, primaries black or black edged with white; secondaries, black on the inner edge and white on the outer edge; remainder pure white, entirely free from ticking. Undercolour: white, bluish-white or light slate, not to be visible when feathers are undisturbed.

Female plumage: As for male except tail feathers are black with beetle-green sheen except the top pair which may or may not be laced with white.

In both sexes: Beak yellow. Eyes rich bay. Face, comb, ear-lobes and wattles red. Legs and feet yellow.

The white

Male and female plumage: Pure snow-white, any straw tinge to be avoided.

In both sexes: Beak yellow. Eyes rich bay. Face, comb, ear-lobes and wattles red. Legs and feet yellow.

Weights

Male	3.40 kg ($7\frac{1}{2}$ lb) min.
Female	2.95 kg ($6\frac{1}{2}$ lb) min.

Scale of points

The barred

Type	30
Barring	20
Colour	15
Size	10
Legs and feet	10
Condition	10
Head	5
	100

The buff

Type (symmetry), shape, size and carriage	30
Colour (general)	20
Quality and texture (general)	15
Condition and fitness	15
Head and comb	10
Eye colour	5
Legs and feet	5
	100

Other varieties

Type	30
Colour	30
Head and eyes	10
Legs and feet	10
Condition	10
Quality and texture	10
	100

Serious defects

The slightest fluff or feather on shanks or feet, or unmistakable signs of feathers having been plucked from them. Legs other than yellow. White in ear-lobes. In the barred, any feathers of any colour foreign to the variety, black feathers excepted; also lopped or rose comb, decidedly wry tail; crooked back, more than four toes, and entire absence of main tail feathers. Other than black feathers in the black. Mealiness, or any black or white in wing or white in tail, spotted hackle, and in the male a spotted saddle and in the female a spotted cushion in the buff. Any coloured feathers in the white. Yellow, straw tinge or sap in white, brown markings, brown undercolour in columbian.

Disqualifications

Trimming, faking and any bodily deformity. Split wing, slipped wing and non-growth of secondaries.

BANTAM

Plymouth Rock bantams are miniatures of their large fowl counterparts and the standards for large fowl should be used. Colours are buff, barred, partridge, blue, black, white, columbian.

Colour

The partridge

Male plumage: Head bright red. Neck hackles: web of feather, solid, lustrous, greenish-black, of moderate width, with a narrow edging of a medium shade of rich brilliant red, uniform in width, extending around point of feather; shaft black; plumage on front of neck black, as clear as possible of red. Wing fronts black; bows a medium shade of rich brilliant red; coverts, lustrous, greenish-black, forming a well-defined bar of this colour across wing when folded; primaries black, lower edges reddish-bay; secondaries black, outside webs reddish-bay, terminating with greenish-black at the end of each feather, the secondaries when folded forming a reddish-bay wing bay between the wing bar and tips of secondary feathers. Back and saddle a medium shade of rich, brilliant red, with lustrous, greenish-black stripe down the middle of each feather, same as in hackle. A slight shafting of rich red is permissible. Tail black; sickles and smaller sickles lustrous, greenish-black; coverts lustrous, greenish-black, edged with a medium shade of rich, brilliant red. Body black; lower feathers slightly tinged with red; fluff black, slightly tinged with red. Breast lustrous black (slight tinge of red allowed). Lower thighs black. Undercolour of all sections slate.

Female plumage: Head deep reddish-bay. Neck reddish-bay, centre portion of feathers black, slightly pencilled, with deep reddish-bay; feathers on front of neck same as breast. Wing shoulders, bows and coverts deep reddish-bay with distinct pencillings of black, outlines of which conform to shape of feathers; primaries black with edging of deep reddish-bay on outer webs; secondaries, inner web black, outer web deep reddish-bay with distinct pencillings of black extending around outer edge of feathers. Back deep reddish-bay with distinct pencillings of black, the outlines of which conform to shape of feathers. Tail black, the two top feathers pencilled with deep reddish-bay on upper edge; coverts deep reddish-bay pencilled with black. Body deep reddish-bay pencilled with black; fluff deep reddish-bay. Breast deep reddish-bay with distinct pencillings of black, the outlines of which conform to shape of feathers. Lower thighs deep reddish-bay pencilled with black. Undercolour slate.

Note: Each feather in back, breast, body, wing bows and thighs to have three or more distinct pencillings.

In both sexes: Beak yellow. Eyes reddish-bay. Legs and feet yellow. Comb, face, wattles and ear-lobes bright red.

The blue

Male and female plumage: One even shade of blue, light to dark, but medium preferred; a clear solid blue, free from mealiness, 'pepper', sandiness or bronze, and quite clear of lacing. Eyes, etc. as in the partridge.

Weights

Male	not to exceed	1.36 kg (3 lb)
Female	not to exceed	1.13 kg ($2\frac{1}{2}$ lb)

Scale of points

The barred

Type	20
Colour	20
Barring	20
Legs and feet	10
Head	5
Tail	5
Size	10
Condition	10
	100

The buff

Type	20
Carriage and size	10
Colour	20
Quality and texture	15
Head	10
Eye colour	5
Legs and feet	5
Condition	15
	100

The partridge

Type	30
Colour and pencilling	30
Head and eyes	10
Legs and feet	10
Condition	10
Quality and texture	10
	100

Serious defects and disqualifications

As for large fowl.

POLAND

LARGE FOWL

Origin: Poland
Classification: Light: Soft feather
Egg colour: White

That the Poland is a very old breed goes without saying, although its ancestry is none too clear. Many connect it with the breed named the Paduan or Patavinian fowl, although this original example is illustrated without muff or beard. Poland (gold or silver spangled, black or white) had a classification at the first poultry show in London in 1845, and was standardized in the first Book of Standards in 1865, with white-crested black, golden and silver varieties included. The white-crested (black or blue) varieties are without muffling, while the others have muffs.

General characteristics: male

Carriage: Sprightly and erect.

Chamois Poland male, bantam

Chamois Poland female, bantam

White-crested black Poland female, bantam

White-crested blue Poland female, bantam

Type: Back fairly long, flat and tapering to the tail. Breast full and round. Flanks deep, shoulders wide, wings large and closely carried. Tail full, neatly spread, and carried rather low, not upright, the sickles and coverts abundant and well curved.
Head: Large, with a decidedly pronounced protuberance on top, and crested. Crest large, full, circular on top and free of any split or parting, high and smooth in front and compact in the centre, falling evenly with long untwisted or reverse-faced feathers far down the nape of the neck, and composed of feathers similar to those of the neck hackle. Beak of medium length, and having large nostrils rising above the curved line of the beak. Eyes large and full. Comb of horn type and very small if any (preference should be given to birds without a comb). Face smooth, without muffling in the white-crested varieties, and completely covered by muffling in the others. Muffling large, full and compact, fitting around to the back of the eyes and almost hiding the face. Ear-lobes very small and round, quite invisible in the muffled varieties. Wattles rather large and long in the white-crested varieties; the others are without wattles.
Neck: Long, with abundant hackle covering the shoulders.
Legs and feet: Legs slender and fairly long, the shanks free of feathers. Toes, four, slender and well spread.

Female

With the exception of the crest, which is globular in shape, the general characteristics are similar to those of the male, allowing for the natural sexual differences.

Colour

The chamois

Male plumage: Buff ground colour with white markings, crest white at the roots and tips, as free as possible of whole white feathers. Muffling mottled or laced, not solid buff. Hackle tipped. Back, saddle and wing coverts distinctly laced or spangled at the tips. Breast, thighs, wing bars and secondaries laced. Primaries tipped, tail coverts and sickles laced, the ends of sickles well splashed.
Female plumage: Except that the primaries are tipped, the colour and markings including the crest are buff ground with white lacing.

In both sexes: Eyes, comb and face red. Ear-lobes blue-white. Beak, legs and feet dark blue and/or horn, soles of feet blue.

The gold

Male and female plumage: As in the chamois, substituting golden-bay as the ground colour with black markings and lacing. Eyes, etc. as in the chamois.

The silver

Male and female plumage: As in the chamois, substituting silver as the ground colour, with black markings and lacing. Eyes, etc. as in the chamois.

White is permissible in the crests of laced varieties over one-year-old.

The self white

Male and female plumage: Pure white throughout. Eyes, etc. as in the chamois.

The self black

Male and female plumage: Rich metallic-black throughout. Eyes, etc. as in the chamois.

The self blue

Male and female plumage: An even shade of blue throughout. Eyes, etc. as in the chamois.

The white-crested black

Male and female plumage: Rich metallic-black, except the crest which is snow-white, with a black band at the base of the crest in front.

In both sexes: Eyes, comb, face and wattles red. Ear-lobes white. Beak, legs and feet blue and/or horn, soles of feet white.

The white-crested blue

Male and female plumage: An even shade of blue throughout, except the crest which is snow-white, with a blue band at the base of the crest in front. Eyes, etc. as in the white-crested black.

The white-crested cuckoo

Male and female plumage: Even banding on all feathers of even shades of grey except the crest, which is snow-white, with a cuckoo band at the base of the crest in front.

In both sexes: Eyes, comb and wattles red. Beak, legs and feet white or blue-white, soles of feet white.

Weights

Male 2.95 kg ($6\frac{1}{2}$ lb)
Female 2.25 kg (5 lb)

Scale of points

The white-crested varieties

Crest	30
Head and wattles	15
Colour	30
Type	5
Size	10
Condition	10
	100

The other varieties

Crest	30
Head and muffling	15
Colour and markings	30
Type	5
Size	10
Condition	10
	100

Serious defects

Split or twisted crest. Comb, if any, other than horn type. Absence of muffling in birds other than the white-crested varieties. Legs other than blue or slate. Other than four toes on each foot. Any deformity. Absence of black, blue or cuckoo in the front of the crests of the white-crested varieties. In the cuckoo solid black or white feather.

BANTAM

Poland bantams follow the standards for their large counterparts.

Weights

Male 680–790 g (24–28 oz)
Female 510–680 g (18–24 oz)

REDCAP

LARGE FOWL

Origin: Great Britain
Classification: Light: Soft feather
Egg colour: White

The Redcap has always been closely associated with Derbyshire and is known as the Derbyshire Redcap. It was a sturdy breed carrying excellent breast meat, and a good egg producer. Farmers used the males freely for crossing to produce layers. As with so many promising utility breeds, the Redcap was bred and exhibited as if its immense comb were all that mattered, head points claiming 45 of the 100 judging points.

General characteristics: male

Carriage: Graceful action and well balanced.
Type: Back broad, moderate in length falling slightly to tail and flat. Breast broad, full, and rounded. Wings moderate in length, neat and fitting closely to the body. Tail full and carried at an angle of about 60°; broad, long and well-arched sickles.
Head: Of medium length and broad. Beak medium in length. Eyes full and prominent. Comb rose with straight leader, full of fine work or spikes, free from hollow in centre, set straight on the head and carried well off the eyes and beak; size about 8.25 × 7.0 cm (3¼ × 2¾ in). Face smooth and of fine texture. Ear-lobes of medium size. Wattles of medium length, well rounded and fine in texture.
Neck: Of moderate length, nicely arched, and with full hackle.
Legs and feet: Legs straight and wide apart. Thighs short and well fleshed, shanks moderately long. Toes, four, well spread.

Female

The general characteristics are similar to those of the male, allowing for the natural sexual differences, with the exception of the comb which is about half the size of the male's. The tail is large and full, and, like that of the male, carried at an angle of about 60°.

Colour

Male plumage: Neck, hackle and saddle to harmonize. Each feather to have a red quill with beetle-green webbing, very finely fringed and tipped with black (the feathers appear to be fringed with red, but if placed on paper they are seen to have a black fringe and tip; the fringe is almost as fine as a hair). Back rich red, tipped with black. Wing bows rich red; Coverts rich red, each feather ending with a black spangle, forming a black bar across the wing; primaries and secondaries black on one side red on the other side, heavily tipped with black. Breast and underparts black. Tail and hangers black.
Female plumage: Hackle as in the male, but nut-brown quill. Back and breast ground colour deep rich nut-brown, free from smuttiness, each feather ending with a half-moon black spangle. The markings on breast, back and wings to be as uniform as possible. Wing primaries and secondaries, as in the male; wing coverts evenly spangled. Tail black.

In both sexes: Beak horn. Eyes red. Comb, face, ear-lobes and wattles bright red. Legs and feet lead colour.

Weight

Cock	2.70–2.95 kg (6–6½ lb);	cockerel	2.50–2.70 kg (5½–6 lb)
Hen	2.25–2.50 kg (5–5½ lb);	pullet	2.00–2.25 kg (4½–5 lb)

Derbyshire Redcap male

Derbyshire Redcap female

Scale of points

Size (style and shape)	10
Tail	5
Head points (comb 25, eyes 5, ear-lobes and wattles 15)	45
Legs and feet	5
Colour	25
Condition	10
	100

Serious defects (for which birds should be passed)

Comb over. White ear-lobes. Round back. Squirrel or wry tail. Feathers on legs. Legs other colour than lead. Other than four toes. Crooked breast. Coarseness. Excessive fat.

RHODE ISLAND RED

LARGE FOWL

Origin: America
Classification: Heavy: Soft feather
Egg colour: Light brown to brown

No breed made such a world progress in so short a time as this American breed. It was developed from Asiatic black-red fowls of Shanghai, Malay and Java types, bred on the farms of Rhode Island Province. Red Javas were known there in 1860, and the original Rhode Island Red had a rose comb, although birds with single combs, probably from brown Leghorn crossings, were bred. They were first exhibited as Rhode Island Reds in 1880 in South Massachusetts. In December 1898, the Rhode Island Red Club of America held their first meeting. In 1904, the Single Comb variety was admitted to the American Poultry Association of Perfection, followed in 1906 by the Rose Combs. The formation of the British Rhode Island Red Club took place in August 1909 and the breed has been one of the most popular in this country for all purposes. Being a gold, males of the breed are utilized extensively in gold-silver sex linked matings.

General characteristics: male

Carriage: Alert, active and well balanced.
Type: Body deep, broad and long. The keel bone long, straight and extending well forward and back, giving the body an oblong look, rather than square. Back broad, long and horizontal, this being modified by rising curve at hackle and a slightly rising curve at the tail coverts. Saddle feathers of medium length and abundant. Tail of medium length, quite well spread, carried fairly well back, increasing the apparent length of the bird. Sickles of medium length, passing a little beyond the main tail feathers. Lesser sickles and tail coverts of medium length and fairly abundant. Breast broad, deep and carried in a line nearly perpendicular with the base of the beak; it should not be carried further back. Fluff moderately full but with the feathers carried fairly close to the body, not Cochin fluff. Wings of good size, well folded, and the flights carried horizontally. With the horizontal back and keel and the breast vertical to the base of the beak, this gives rise to the breed being described as 'brick shaped'.
Head: Of medium size, carried horizontally and slightly forward. Beak medium in length and slightly curved. Eyes full bright and prominent. Comb single or rose. The single of medium size, fine texture, set firmly in the head, perfectly straight and upright, with five even and well-defined serrations, those in front and rear smaller than the centre ones, of

Rhode Island Red male, bantam

Rhode Island Red female, bantam

considerable breadth where it is fixed to the head. The rose of medium size, low, set firmly on the head, the top oval in shape, and the surface covered with small points, terminating in a small spike at the rear. The comb to conform to the general curve of the head. Face smooth and of fine texture. Ear-lobes fairly well developed. Wattles medium and equal in length, moderately rounded and of fine texture.
Neck: Of medium length, carried slightly forward, and covered with abundant hackle, flowing over the shoulders but not too loosely feathered.
Legs and feet: Legs well apart. Thighs large, of medium length and well covered with feathers. Shanks of medium length, well rounded and free from feathers. Toes, four, of medium length, straight, strong and well spread.

Female

The general characteristics are as those of the males, apart from the sexual differences, e.g. smaller wattles, comb and no sickle feathers, leaving the tail feathers exposed. The tail, however, should not form an apparent angle with the back, nor must it be met by a high rising cushion. It should be quite well spread, carried fairly well back increasing the apparent length of the bird. Neck hackle should be sufficient, but not too coarse in feather. In the mature female the back would be described as broad, while in the pullet it would look somewhat narrower in proportion to the length of her body. The curve from the horizontal back to the hackle or tail should be moderate and gradual.

Colour

Male plumage: The neck red, harmonizing with back and breast. Wing primaries, the lower web black, with red along outer edging permissible and the upper web red; secondaries, the lower web red and the upper black; flight coverts black; wing bows and coverts red. Tail, main feathers, including the sickles, black or greenish-black; coverts mainly black, but they may become russet or red as they approach the saddle. The hackle to show a rich brilliant red plumage with no black ticking or lacing. The general surface of the plumage should be a rich brilliant red, except where black is specified. It should be free from shafting, mealy appearance or brassy effect. Absolute evenness of colour is desirable. The bird should be so brilliant in lustre as to have a glossed appearance. The undercolour and quill feather should be red or salmon. With the saddle parted, showing the undercolour at the base of the tail, appearance should be red or salmon, not whitish or smoky. Black or white in the undercolour of any section is undesirable.
Female plumage: Neck hackle to be red, the tips of the lower feather may show black ticking, but not heavy lacing. The tail should be black or greenish-black, however, the upper webs of the two main tail feathers may be edged with red. In all sections of the wing the undercolour and quills of the feathers are as in the males. With the remainder of the plumage the surface should be a rich dark, even and lustrous red, but not as brilliant a lustre as in the male. It should be free from shafting or mealy appearance. (*Note:* The 'red' colour nowadays favoured is an extremely deep chocolate-red and, though some breeders disagree with this description, few birds of lighter colour receive prizes.)

In both sexes: Other things being equal, the specimen having the richest undercolour shall receive the award. Beak red-horn or yellow. Eyes red. Face, comb, wattles and ear-lobes bright red. Legs and feet yellow or red-horn.

Weights

Cock	3.85 kg ($8\frac{1}{2}$ lb);	cockerel	3.60 kg (8 lb)
Hen	2.95 kg ($6\frac{1}{2}$ lb);	pullet	2.50 kg ($5\frac{1}{2}$ lb)

Scale of points

Type (shape 10, size 10, carriage and symmetry 10)	30
Colour (general)	25
Quality and texture (10 + 10)	20
Head and comb (5 + 5)	10
Eye colour	10
Legs	5
	100

Serious defects

Feather or down on shanks or feet, or unmistakable indications of a feather having been plucked from them. Badly lopped comb, side sprig or sprigs on the single comb. Other than four toes. Entire absence of main tail feathers. An ear-lobe showing white (this does not mean pale ear-lobes, but enamelled white.) Two absolutely white (so-called wall or fish) eyes. Squirrel or wry tail. A feather entirely white that shows in the outer plumage. Diseased specimens, crooked backs, deformed beaks, shanks and feet other than yellow or red-horn colour. Birds showing any deformity. A pendulous crop shall be cut hard. Coarseness. Toes not straight and well spread. Super-fineness or frizzled, Silkie type defective feather often developed through concentration on lustrous dark plumage. Under all disqualifying causes, the specimen shall have the benefit of the doubt. Robustness is of vital importance.

BANTAM

Rhode Island Red bantams are miniatures of their large fowl counterparts and the standard for those should be followed.

Weights

Male	790–910 g (28–32 oz)
Female	680–790 g (24–28 oz)

ROSECOMB BANTAM

Origin: Great Britain
Classification: True bantam
Egg colour: White or cream

The Rosecomb bantam is a gem of show birds. In former days it achieved probably the highest pitch of artificial perfection ever achieved in exhibition birds.

General characteristics: male

Carriage: Cobby but not dumpy. The back should show one sweeping curve from neck to sickles.
Type: Body short and broad. Back short, shoulders broad and flat. Breast carried well up and forward, with a bold curve from wing bow to wing bow. Wings carried rather low, slowing only front half of thighs. Wide flight feathers round ended and broad to ends. Stern flat, broad and thick (not running off to nothing at setting-on of tail), with abundant feather; the saddle hackle long and plentiful and extending from tail to middle of back. Tail carried well back, main feathers broad and overlapping neatly; the sickles being long, circled with a bold sweep, broad from base to rounded ends, main tail feathers not projecting beyond the sickles. Furnishing feathers plentiful, broad from base to end, round ended and uniformly curved with the sickles but hanging somewhat shorter; side

Black Rosecomb male

Black Rosecomb female

hangers broad and long and with the hackles filling the space between stern and wing ends. All feather broad to ends.
Head: Short and broad. Beak stout and short. Comb rose, neat and long, with square well-filled front, set firmly, tapering to the setting-on of the spike or leader; top perfectly level and crowded with small round spikes. The leader stout at base, firm, long and perfectly straight, tapering to a fine point. Comb and leader rise slightly from front to rear in one line. Face of fine texture. Ear-lobes absolutely round, with rounded edges, of uniform thickness all over, not hollow or dished, firmly set on the face and kid-like in texture; not smaller than 1.88 cm ($\frac{3}{4}$ in) or larger than 2.19 cm ($\frac{7}{8}$ in). Wattles round, neat and fine.
Neck: Rather short, well curved, with wide feathers, the hackle falling gracefully and plentifully over shoulders and wing bows and almost reaching the tail.
Legs and feet: Legs short. Thighs set well apart, stout at top and tapering to hocks. Shanks rather short, round, fine and free of feathers. Toes, four, straight and well spread.

Female

With the exception of the ear-lobes, which should not be larger than the now defunct silver threepenny piece (approx. 1.56 cm ($\frac{5}{8}$ in)), and the wings, which are not carried so low but are tucked up, the general characteristics are similar to those of the male, allowing for the natural sexual differences.

(*Note:* Standard sizes of ear-lobes are usually considerably exceeded in show specimens.)

Colour

The black

Male and female plumage: Black with brilliant green sheen from head to end of tail, the wing bar with extra bright green sheen. Tail feathers and sickles to be rich in green sheen.
In both sexes: Beak black, eyes hazel or brown. Legs and feet black.

The blue

Male and female: Blue of medium shade, free from lacing. The plumage of hackles, back and shoulders in males of a darker shade.

The white

Male and female plumage: Snow-white, free from straw tinge.
In both sexes: Bleak white, eyes red. Legs and feet white.

In both sexes and all colours

Comb, face and wattles brilliant cherry-red. Ear-lobes spotlessly white, especially near wattles.

Weights

Male	570–620 g (20–22 oz)
Female	450–510 g (16–18 oz)

Scale of points

Head (comb 20, lobes 15)	35
Tail	15
Colour	12
Type	15
Condition	15
Legs, etc.	8
	100

Serious defects

Stiltiness. Narrow chest or back. Hollow-fronted or leafy comb. Coarse bone. Tightly carried wings. Narrow feathers. Blushed lobes. Coloured feathers. White in face. In blacks, grizzled or brown flights, purple sheen or barring, light legs.

RUMPLESS GAME

LARGE FOWL

Origin: Great Britain
Classification: Hard feather: Rare
Egg colour: Tinted

As is common with any rumpless breed, the parson's nose or 'Caudal Appendage' (uropygium) is missing. Foreign breeds like the Barbu d'Anvers has a rumpless version called Barbu du Grubbe. The Barbu d'Uccle has the Barbu d'Everberg as its rumpless version and, even in Japan, there is a Rumpless Yokohama, funny though it may sound, with its long saddle hackle making it feasible. A genetic accident with Old English Game many years ago probably created our own Rumpless Game. The breed, though popular in the bantam form is not often seen in large fowl.

General characteristics: male

Carriage: Bold and upright.
Type: Body short and small. Back steeply sloping. Breast full and prominent. Wings large and long. Tail completely absent, the whole of the lower back being covered with the saddle feather.
Head: Small and tapering. Beak rather large. Eyes large and bold. Comb single, small and thin, upright. Ear-lobes and wattles fine, small and thin.
Neck: Long and upright; hackle close and wiry.
Legs and feet: Legs strong and of medium length. Thighs short, round and muscular. Toes, four, long, straight and tapering.
Plumage: Hard and close.

Female

The general characteristics are similar to those of the male, allowing for the natural sexual differences.

Colour

Plumage colour is of secondary importance in this breed, and almost any recognized 'Game' colour is acceptable. Face, comb, wattles and ear-lobes should be bright red.

Weights

Male 2.25–2.70 kg (5–6 lb)
Female 1.80–2.25 kg (4–5 lb)

Scale of points

Type	25
Carriage and bearing	20
Head	15
Colour	15
Legs and feet	5
Size	15
Condition	5
	100

Rumpless Game male, bantam

Disqualifications

Any sign of tail. Other than single comb. Any deformity.

BANTAM

Rumpless Game bantams should follow exactly the large fowl standard.

Weights

Male 620–740 g (22–26 oz)
Female 510–620 g (18–22 oz)

SCOTS DUMPY

LARGE FOWL

Origin: Great Britain
Classification: Light: Soft feather
Egg colour: White

This breed has been bred in Scotland for more than a hundred years, and the birds used to be known also as Bakies, Crawlers and Creepers. Fowls having identical dumpy characters have been shown to exist as early as AD 900. The breed is considered an ideal sitter and mother.

General characteristics: male

Carriage: Heavy, with a waddling gait, the extreme shortness of its legs giving the bird

Black Scots Dumpy male, large

Black Scots Dumpy female, large

the appearance of 'swimming on dry land'. Shortness of leg alone should not constitute the breed's claim to notice. The large, low, heavy body, and other points of excellence must be possessed also.
Type: Body long and broad. Back broad and flat. Breast deep. Wings of medium size and neatly carried. Tail full and flowing, the sickles well arched.
Head: Fine. Beak strong and well curved. Eyes large and clear. Comb single, of medium size, upright and straight, free from side sprigs, and the back following the line of the skull, evenly serrated on top. Face smooth. Ear-lobes small and close to the neck. Wattles of medium size.
Neck: Of fair length, in keeping with the size of the body, and covered with flowing hackle.
Legs and feet: Legs very short, the shanks not exceeding 3.75 cm ($1\frac{1}{2}$ in). Toes, four, well spread.

Female

The general characteristics are similar to those of the male, allowing for the natural sexual differences.

Colour

Male and female plumage: There is no fixed plumage colour, but the varieties most widely seen are black, cuckoo, white, brown, gold and silver. Cuckoo to be light grey with dark grey bands.

In both sexes: Eyes red. Comb, face, wattles and ear-lobes bright red. Beak, legs and feet white, except in the black variety where they should be black or slate, and in the cuckoo mottled.

Weights

Male	3.20 kg (7 lb)
Female	2.70 kg (6 lb)

Scale of points

Type	40
Size	20
Head	15
Condition	15
Colour	10
	100

Serious defects

White ear-lobes. Yellow or feathered shanks or feet. Long legs. Any deformity.

BANTAM

Scots Dumpy bantams should follow the large fowl standard.

Weights

Male	800 g ($1\frac{3}{4}$ lb)
Female	675 g ($1\frac{1}{2}$ lb)

SCOTS GREY

LARGE FOWL

Origin: Great Britain
Classification: Light: Soft feather
Egg colour: White

A light, non-sitting breed originated in Scotland, it has not been bred extensively outside that country where, even if it is less popular today, it will doubtless be maintained by keen breeders. It has been bred there for over two hundred years.

General characteristics: male

Carriage: Erect, active and bold.
Type: Body compact, full of substance and fairly long. Back broad and flat. Breast deep, full and carried upwards. Wings moderately long and well tucked, the bow and tip covered by the neck and saddle hackles. Tail fairly long and well up (but not squirrel fashion) with full sickles.
Head: Long and fine. Beak strong and well curved. Eyes large and bright. Comb single, upright, of medium size, with well-defined serrations, the back following the line of the skull. Face of fine texture. Ear-lobes of medium size. Wattles of medium length with a well-rounded lower edge.
Neck: Finely tapered and with profuse hackle flowing on the back and shoulders.
Legs and feet: Legs long and strong. Thighs wide apart but not quite as prominent as those of Game fowl. Shanks free from feathers. Toes, four, straight and spreading, stout and strong.
Handling: Firm, and somewhat similar to the Game fowl.

Female

With the exception of the comb, either erect or falling slightly over, the general characteristics are similar to those of the male, allowing for the natural sexual differences.

Colour

Male plumage: Barred. Ground colour of body, thighs and wing feathers steel-grey. The barring is black with a metallic lustre, that of the body, thighs and wing feathers straight across, but that of the neck hackle, saddle, and tail slightly angled or V-shaped. The alternating bars of black are of equal width and proportioned to the size of the feather. The bird should 'read' throughout, i.e. the shade should be the same from head to tail. The plumage should be free from red, black, white, or yellow feathers, and the hackle, saddle, and tail should be distinctly and evenly barred, while the markings all over should be rather small, even, and sharply defined.
Female plumage: Similar to that of the male, except that the markings are not as small, and produce an appearance somewhat resembling a shepherd's tartan.

In both sexes: Beak white or white streaked with black. Eyes amber. Comb, face, wattles and ear-lobes bright red. Legs and feet white or white mottled with black, but not sooty.

Weights

Male 3.20 kg (7 lb)
Female 2.25 kg (5 lb)

Scots Grey male, bantam

Scots Grey female, bantam

Scale of points

Colour and markings	30
Size	10
Type	30
Head	10
Condition	10
Legs and feet	10
	100

Serious defects

Any bodily deformity. Any characteristic of any other breed not applicable to the Scots Grey.

BANTAM

Scots Grey bantams follow the large fowl standard.

Weights

Male 620–680 g (22–24 oz)
Female 510–570 g (18–20 oz)

SEBRIGHT BANTAM

Origin: Great Britain
Classification: True bantam
Egg colour: White or cream

This breed is a genuine bantam and one of the oldest British varieties. It has no counterpart in large breeds, but has played a part in the production of other laced fowl, notably Wyandottes. There are two colours, gold and silver.

General characteristics: male

Carriage: Strutting and tremulous, on tip-toe, somewhat resembling a fantail pigeon.
Type: Body compact, with broad and prominent breast. Back very short. Wings large and carried low. Tail square, well spread and carried high. Sebright males are hen feathered, without curved sickles or pointed neck and saddle hackles.
Head: Small. Beak short and slightly curved. Comb rose, square fronted, firmly and evenly set on, top covered with fine points, free from hollows, narrowing behind to a distinct spike or leader, turned slightly upwards. Eyes full. Face smooth. Ear-lobes flat, and unfolded. Wattles well rounded.
Neck: Tapering, arched and carried well back.
Legs and feet: Legs short and well apart. Shanks slender and free from feathers. Toes, four, straight and well spread.
Plumage: Short and tight, feathers not too wide but never pointed. (Almond-shaped feather is desired.)

Female

The general characteristics are similar to those of the male, allowing for the natural sexual differences. Her neck is upright.

Colour

The gold

Male and female plumage: Uniform golden-bay with glossy green-black lacing and

Silver Sebright male

Gold Sebright female

dark undercolour. Each feather evenly and sharply laced all round its edge with a narrow margin of black. Shaftiness is undesirable.

The silver

Male and female plumage: Similarly marked on pure, clear silver-white ground colour.

In both sexes and colours

Beak dark horn in golds; dark blue or horn in silvers. Eyes black, or as dark as possible. Comb, face, wattles and ear-lobes dark purple or dull red (mulberry). Legs and feet slate-blue.

Although in males the purple or mulberry face is seldom obtainable, the eye should be dark and surrounded with a dark cere.

Weights

Male	620 g (22 oz)
Female	510 g (18 oz)

Scale of points

Lacing	25
Comb	5
Face and lobes	10
Ground colour	15
Tail	10
Type	20
Weight	5
Condition	10
	100

Note: There is at present a decided move to improve type and to discourage the prevailing whip tails and narrow build, particularly in females.

Serious defects

Single comb. Sickle feathers or pointed hackles on the male. Feathers on shanks. Legs other than slate-blue. Other than four toes. Any deformity.

SHAMO

LARGE FOWL

Origin: Asia
Classification: Hard feather: Rare
Egg colour: White or tinted

The Shamo is a Japanese bird of Malay type, originally imported to Japan from Thailand in the seventeenth century – the name being a corruption of Siam, the old name for Thailand. In Japan it was developed into a fighting bird of unmatched courage and ferocity. Its feathers are sparse but strong and shiny, and its powerful bone structure and well-muscled body and legs, coupled with its erect posture, make it an impressive and striking bird. In Japanese classification, birds are divided into O-Shamo (large) and Chu-Shamo (medium) but in this country, since its importation in the early 1970s, the term 'Shamo' covers both sizes of large fowl.

General characteristics: male

Carriage: Upright and proud.

Shamo male

Shamo female

Type: General appearance fierce and powerful. Large firm body. Broad full breast with deep keel. Back long and flat, broadest at shoulders, sloping down towards tail and gradually tapering from upper side of thigh. Abdomen firm and well-muscled. Wings short, big, strong and bony, carried close to body, not showing on the back but with prominent shoulders. Tail carried below horizontal.
Head: Deep and broad with wattles and ear-lobes small or absent. Beak strong and broad. Eyes deep set under overhanging brows. Comb triple.
Neck: Round, long, strong, slightly curved but almost erect.
Legs and feet: Thighs long, round and muscular. Legs medium to long, thick and strong with slight bend at hock. Toes four, long and well spread.
Plumage: Feathers short, narrow, hard and brilliant. Scant, or bare showing red skin at throat, keel and point of wing. Neck hackle feathers permitted to curl towards back of neck.
Handling: Firm fleshed and muscular.

Female

Generally of a less upright stance than the male – otherwise the general characteristics are similar, allowing for the natural sexual differences.

Colour

Male varieties: Red-hackled black, dark red-hackled black, yellow-hackle black, ginger, white, black, splash, blue.
Female varieties: Wheaten, partridge, ginger, white, black, splash, blue.

In both sexes and all colours

Beak yellow or horn. Legs and feet yellow or black with yellow undercolour. Comb, face, throat and ear-lobes brilliant red. Eyes very light to orange.

Weights

Male	3.4 kg ($7\frac{1}{2}$ lb) min.
Female	2.5 kg ($5\frac{1}{2}$ lb) min.

Scale of points

Type and carriage	40
Head	20
Feather/condition	20
Colour	10
Legs and feet	10
	100

Serious defects

Poor carriage. Wry tail.

SICILIAN BUTTERCUP

LARGE FOWL

Origin: Europe
Classification: Light: Rare
Egg colour: White

The Sicilian Buttercup was first imported into this country in 1912 by Mrs Colbeck of Yorkshire. There is never any mistaking this breed because of its distinctive saucer-shaped cup comb. The Buttercup was plentiful in the 1960s when many were used for laying trials because of their ability to produce many eggs from little food. The Sicilian Flowerbird which was known in the brown variety and was a smaller bird, now no longer exists. The Sicilian Buttercup now exists in the gold and silver form and is not known in the golden duckwing or white.

General characteristics: male

Carriage: Upright, bold and active.
Type: Body moderately long and deep; broad shoulders and narrow saddle. Full round breast. Broad back sloping downwards to the saddle, which rises in a slightly concave outline to the base of the tail. Long wings, closely tucked. Fairly large tail with long main feathers, carried at an angle of 45°, and fitted with well-curved sickles and abundant coverts.
Head: Skull long and deep. Beak of medium length. Eyes full and keen. Comb beginning at the base of the beak with a single leader and joined to a cup-shaped crown, set firmly on the centre of the skull and surmounted with well-defined and regular points, of medium size and fine texture, and free from decided spikes in the cavity or centre. Ear-lobes almond-shaped, flat, smooth, and close fitting. Wattles thin and well rounded.
Neck: Rather long, with hackle flowing well over the shoulders.
Legs and feet: Of moderate size and length. Thighs well apart. Shanks slender and free of feathers. Toes, four, straight and spreading.

Female

With the exception of the comb (smaller and lower in proportion) the general characteristics are similar to those of the male, allowing for the natural sexual differences.

Sicilian Buttercup large fowl
Male

Colour

The golden

Male plumage: Neck hackle, back, saddle, shoulders and wing bows bright lustrous orange-red. Cape (at base of neck) dark buff marked with distinct black spangles and covered by hackle. Wing bar and bay an even shade of red-bay; primaries black, lower web edged with bay; secondaries red-bay on outer web, black on inner. Breast red-bay. Body light bay. Fluff rich bay shading to light bay on stern, and some feathers on the body fluff with distinct black spangles. Tail black, sickles and coverts green-black, the former showing red-bay at their base and the coverts edged with that colour.

Female plumage: Hackle lustrous golden-buff. Breast and thighs light golden-buff, plain from throat to middle of breast, elsewhere with black spangles. Tail dull black, except the two highest feathers mottled with buff. Wing bow and bar golden-buff with parallel rows of elongated black spangles, each spangle extending slightly diagonally across the web; quill and edge of feathers golden-buff; primaries black edged with buff; secondaries golden-buff regularly barred with black on outer web, black on inner. Back golden-buff regularly spangled with black (the same pattern as the wing bow) and extending over the entire surface, including the saddle and the tail coverts.

The silver

Male and female: Except that the ground is silver-white (free from yellow or straw tinge) instead of red or bay, similar to the golden.

In both sexes and colours

Beak dark horn lightly shaded with yellow. Eyes red-bay. Comb, face, wattles and ear-lobes bright red (more than one-third white in lobes a serious defect). Legs and feet willow green.

Weights

Male 2.95 kg ($6\frac{1}{2}$ lb)
Female 2.50 kg ($5\frac{1}{2}$ lb)

Scale of points

Head	30
Colour	25
Type	20
Size	10
Legs and feet	10
Condition	5
	100

Serious defects

Spikes more than 2.5 cm (1 in) long in cup (or cavity of comb) of male, or any indication of spike in cup (or cavity of comb) of female. Solid white ear-lobes. Shanks other than green. Feathers or stubs on shanks or toes. Any deformity. In golden, solid white in any part of the plumage (except undercolour) or black striping in the male's hackles.

BANTAM

Sicilian Buttercup bantams should follow exactly the large fowl standard.

Weights

Male 735 g (26 oz)
Female 620 g (22 oz)

SILKIE

LARGE FOWL

Origin: Asia
Classification: Light: Soft feather
Egg colour: Tinted or cream

Silkie fowls have been mentioned by authorities for several hundred years, although some think they originated in India, while others favour China and Japan. Despite light weights the Silkie is not regarded as a bantam in this country but as a large fowl light breed, and as such it must be exhibited. Its persistent broodiness is a breed characteristic, and either pure or crossed, the breed provides reliable broodies for the eggs of large fowl or bantams.

General characteristics: male

Carriage: Stylish, compact and lively.
Type: Body broad and stout looking. Back short, saddle silky and rising to the tail, stern broad and abundantly covered with fine fluff, saddle hackles soft, abundant, and flowing. Breast broad and full. Shoulders stout, square, and fairly covered with neck hackle. Wings soft and fluffy at the shoulders, the ends of the flights ragged and 'osprey plumage' (i.e. some strands of the flight hanging loosely downward). Tail short and very ragged at the

White Silkie male, large

White Silkie female, large

Gold Silkie male, large

Black Silkie female, large

end of the harder feathers of the tail proper. It should not be flowing, but forming a short round curve.
Head: Short and neat, with good crest, soft and full, as upright as the comb will permit, and having half a dozen to a dozen soft, silky feathers streaming, gracefully backwards from lower back part of crest, to a length of about 3.75 cm ($1\frac{1}{2}$ in). The crest proper should not show any hardness of feather. Beak short and broad at base. Eyes brilliant black and not too prominent. Comb almost circular in shape, preferably broader than long, with a number of small prominences over it and having a slight indentation or furrow transversely across the middle. Face smooth. Ear-lobes more oval than round. Wattles concave, nearly semi-circular, not long or pendant.
Neck: Short or medium length, broad and full at base with the hackle abundant and flowing.
Legs and feet: Free from scaliness. Thighs wide apart and legs short. No hard feathers on the hocks but a profusion of soft silky plumage is admissible. Thighs covered with abundant fluff. The feathers on the legs should be moderate in quantity. Toes five in number, the fourth and fifth diverging from one another. The middle and outer toes feathered, but these feathers should not be too hard.
Plumage: Very silky and fluffy with a profusion of hair-like feathers.

Female

Saddle broad and well cushioned with the silkiest of plumage which should nearly smother the small tail, the ragged ends alone protruding, and inclined to be 'Cochiny' in appearance. The legs are particularly short in the female, and the underfluff and thigh fluff should nearly meet the ground. The head crest is short and neat, like a powder puff, with no hard feathers, nor should the eye be hidden by the crest, which should stand up and out, not split by the comb. Ear-lobes small and roundish. Wattles either absent or very small and oval in shape. Comb small. Other characteristics are similar to those of the male, allowing for the natural sexual differences.

Colour

The black

Male and female plumage: Black all over with a green sheen in the males. A minimal amount of colour in hackle is permissible, but not desirable.

The blue

Male and female plumage: An even shade of blue from head to tail.

The gold

Male and female plumage: A bright even shade of gold throughout, with darker feathers permissible in the tails of both sexes.

The white

Male and female plumage: Snow-white.

The partridge

Male plumage: Head and crest dark orange. Hackles orange/yellow, free from washiness, each feather having a clear black stripe down the centre. Back and shoulders dark orange. Wing bar, solid black; primaries black, free from any white; secondaries, outer web dark orange, inner web black, the dark orange alone showing when the wing is closed. Tail and sickles black. Leg and foot feather black. Breast and fluff black. Undercolour slate-grey, free from white.
Female plumage: Neck and breast lemon striped black. Hackle feathers black centre with lemon edge. Crest, lemon and black mingling. Body, including wings and cushion, black barring on soft partridge brown. Undercolour slate-grey. Leg foot feather colour as the body. Black permissible in the tail.

Colour

Male and female plumage: Black with a beetle-green sheen, and free of purple bars.
In both sexes: Beak dark horn. Eyes black. Comb and wattles bright red. Face and ear-lobes white. Legs and feet pale slate.

Weights

Male 3.20 kg (7 lb)
Female 2.70 kg (6 lb)

Scale of points

Face and lobes	35
Comb and wattles	15
Type	15
Size	15
Colour	10
Condition	10
	100

Serious defects

Blue, pink or red in face or lobes. Coarse 'cauliflower' face or lobes. Male's comb not erect, side sprigs on comb. Lobes pointed at the bottom. Black or dark legs or feet. Any deformity.

BANTAM

Bantam white-faced Spanish should follow exactly the large fowl standard.

Weights

Male 1075 g (38 oz)
Female 910 g (32 oz)

SULMTALER

LARGE FOWL

Origin: Austria
Classification: Heavy: Rare
Egg colour: Cream to light brown

The origin of the Sulmtaler lies south and south-west of Graz, capital of the Austrian county Stiermarken. Especially in the valleys of Kainach, Lassnitz, Sulm and Saggau (tal = valley), heavy fowls were bred for high quality fattening, mainly being fed on locally grown maize. From 1865 to 1875, these birds were crossed with Cochin, Houdan and Dorkings and then crossed back again to the local fowls from Stiermarken. By 1900, the Sulmtaler had been developed as a breed in its own right and spread into Germany, Holland and England. It is a hardy fowl, fast growing, easy to fatten and a good utility breed.

General characteristics: male

Type: Body full and deep, ratio of deepness to broadness 3:2. Back broad and almost horizontal, medium length with a full saddle and no development of a cushion. Breast very deep, broad, full and well rounded. Wings medium length carried closely. Abdomen

Sulmtaler male, bantam

Sulmtaler female, bantam

Sultan male

Sultan female

Serious defects

Any deformity. Coloured plumage. Toes other than five in number.

BANTAM

Sultan bantams should follow exactly the large fowl standard.

Weights

Male	680–790 g (24–28 oz)
Female	510–680 g (18–24 oz)

SUMATRA

LARGE FOWL

Origin: Asia
Classification: Light: Rare
Egg colour: White

The Sumatra, which comes from the island of Sumatra or the Malay Archipelago, was admitted to the American standard in 1883. With the help of Lewis Wright and Frederick R. Eaton the British standard was drawn up in 1906 under the name of Black Sumatras. A long, flowing tail, carried horizontally, and a pheasant-like carriage are distinguishing characteristics. Sumatras are prolific layers of white eggs and excellent sitters, especially being used to hatch waterfowl. In the late 1970s, a strain of bantams was recreated.

General characteristics: male

Carriage: Straight and upright in front, pheasant-like, giving a proud and stately appearance.
Type: Body rather long, very firm and muscular, broad, full and rounded breast. Back of medium length, broad at shoulders, very slightly tapering to tail. Saddle hackle very long and flowing. Stern narrower than shoulders, but firm and compact. Strong, long and large wings, carried with fronts lightly raised, the feathers folded very closely together, not carried drooping or over the back. Long drooping tail with a large quantity of sickles and coverts, which should rise slightly above the stern and then fall streaming behind, nearly to the ground. Sickle and covert feathers not too broad.
Head: Skull small, fine, and somewhat rounded. Beak strong, of medium length, slightly curved. Eyes large and very bright, with a quick and fearless expression. Comb pea, low in front, fitting closely, the smaller the better. Face smooth and of fine texture. Ear-lobes as small as possible and fitting very closely.
Neck: Rather long, and covered with very long and flowing hackle.
Legs and feet: Of strictly medium length, thick and strong. Thighs muscular, set well apart. Shanks straight and strong, set well apart, with smooth, even scales not flat or thin. (*Note:* There is no objection to two or more spurs on each leg, it being a peculiarity of the breed for this to occur.) Feet broad and flat. Toes, four, long, straight, spread well apart, with strong nails, the back toe standing well backward and flat on the ground.
Plumage: Very full and flowing, but not too soft or fluffy.

Female

Main tail feathers are wide and well spread, the top two feathers curved in a convex manner and carried nearly horizontally. Coverts are moderately long, wide and abundant. Otherwise the general characteristics are similar to those of the male, allowing for the natural sexual differences.

Black Sumatra male, bantam

Black Sumatra female, large

Colour

The black

Male and female plumage: Very rich beetle-green (green-black) with as much sheen as possible.

The blue

Male plumage: Hackles, saddle, wing bow, back and tail very dark slate-blue. Remainder medium slate-blue, each feather to show lacing of darker shade as on the back.
Female plumage: Medium slate-blue, laced with darker shade throughout, except head and neck, a dark slate-blue.

In both sexes and colours

Beak black. Eyes very dark brown or black (black preferred). Face, comb, ear-lobes and throat black or gypsy faced (black preferred). Legs and feet dark olive or black (black preferred).

Weights

Male 2.25–2.70 kg (5–6 lb)
Female 1.80 kg (4–5 lb)

Scale of points

Type	20
Head (beak 5, eyes 5, other points 10)	20
Colour	15
Feather, quantity of	15
Condition	15
Legs and feet	10
Neck	5
	100

Serious defects

Single or rose comb. Any sign of dubbing. Red colour in comb, face or throat. Any sign of wattles. Other than four toes. Any deformity.

BANTAM

Sumatra bantams should follow exactly the large fowl standard.

Weights

Male 735 g (26 oz)
Female 625 g (22 oz)

SUSSEX

LARGE FOWL

Origin: Great Britain
Classification: Heavy: Soft feather
Egg colour: Tinted

This is a very old breed, for although we do not find it included in the first Book of Standards of 1865, at the first poultry show of 1845 the classification included Old Sussex or Kent fowls, Surrey fowls and Dorkings. The oldest variety of the Sussex is the speckled. Brahma, Cochin and silver grey Dorking were used in the make-up of the light. The

Speckled Sussex female, large

Light Sussex male, bantam

Buff Sussex female, bantam

Silver Sussex female, bantam

earlier reds had black breasts, until the red and brown became separate varieties. Old English Game has figured in the make-up of some strains of browns. Buffs appeared about 1920, clearly obtained by sex-linkage within the breed. Whites came a few years later, as sports from lights. Silvers are the latest variety. The light is the most widely kept in this country today among standard as well as commercial breeders. It is one of our most popular breeds for producing table birds. At the time when sex-linkage held considerable popularity the light Sussex was one of the most popular breeds of the day, the females being in considerable demand for mating to gold males. At an even earlier stage, the Sussex breed formed the mainstay of the table poultry market in and around the Heathfield area. The Sussex Breed Club was formed as far back as in 1903 and is now one of the oldest breed clubs in Britain.

General characteristics: male

Carriage: Graceful, showing length of back, vigorous and well balanced.
Type: Back broad and flat. Breast broad and square, carried well forward, with long, straight and deep breast bone. Shoulders wide. Wings carried close to the body. Skin clear and of fine texture. Tail moderate size, carried at an angle of 45°.
Head: Of medium size and fine quality. Beak short and curved. Eyes prominent, full and bright. Comb single, of medium size, evenly serrated and erect, and fitting close to the head. Face smooth and of good texture. Ear-lobes and wattles of medium size and fine texture.
Neck: Gracefully curved with fairly full hackle.
Legs and feet: Thighs short and stout. Shanks short and strong, and rather wide apart, free from feather, with close-fitting scales. Toes, four, straight and well spread.
Plumage: Close and free from any unnecessary fluff.

Female

The general characteristics are similar to those of the male, allowing for the usual sexual differences.

Colour

The brown

Male plumage: Head and neck hackles rich dark mahogany striped with black. Saddle hackle same as neck hackle. Back and wing bow rich dark mahogany. Wing coverts forming the bar blue-black; secondaries and flights black, edged with brown. Breast, tail and thighs black.
Female plumage: Head and neck hackles brown striped with black. Back and wings dark brown, finely peppered with black. Breast and underbody clear pale wheaten-brown. Flights black, edged with brown. Tail black.

Buff

Male and female plumage: Body rich even golden-buff. Head and neck hackles buff, sharply striped with green-black. Wings buff, with black in the flights. Tail and coverts greeny black. Dark in undercolour, not penalized at present, but buff is desirable.

The light

Male and female plumage: Head and neck hackles white, striped with black, the black centre of each feather to be entirely surrounded by a white margin. Wings white, with black in flights. Tail and coverts black. Remainder pure white throughout.

The red

Male and female plumage: Head and neck hackles rich dark red, striped with black. Body and wing bow rich dark red, one uniform shade throughout free from pepperiness. Wings rich dark red with black in the flights. Tail black; coverts rich dark red. Undercolour slate.

The speckled

Male plumage: Head and neck hackles rich dark mahogany, striped with black and tipped with white. Wing bow speckled; primaries white, brown and black. Saddle hackle similar to neck hackle. Tail, main feathers black and white, sickles black with white tips. Remainder rich dark mahogany, each feather tipped with a small white spot, a narrow glossy black bar dividing the white from the remainder of the feather. Undercolour slate and red with a minimum of white.

Female plumage: Head, neck and body ground colour rich dark mahogany, each feather tipped with a small white spot, a narrow glossy black bar dividing the white from the remainder of feathers, the mahogany part of feather free from pepperiness, neither of the colours to run into each other, and to show the three colours distinctly; undercolour as for male. Tail black and brown with white tip. Flights black, brown and white.

The silver

Male plumage: Head, neck and saddle hackles white striped with black, the black centre of each feather to be entirely surrounded by a white margin. Wing bow and back silvery white; coverts forming bar black; flights and secondaries black tinged with grey. Breast black with white shafts, and silver lacing round feathers. Thighs dark grey showing faint lacing. Tail black. Undercolour grey-black shading to white at skin.

Female plumage: Head and neck hackles as in the male. Back and wing bow greyish, each feather showing white shaft with fine silver lacing surrounding it; flights and secondaries greyish-black. Tail black. Breast and thighs lighter shade of greyish-black with white shafts and silver lacing to correspond with the top colour. Undercolour as in the male.

The white

Male and female plumage: Pure white throughout and to the skin.

In both sexes and all colours

Beak white or horn generally, dark or horn with the brown and dark shading to white with the silver. Eyes: brown – brown or red; buff, red and speckled – red; light, silver and white – orange. Face, comb ear-lobes and wattles red. Shanks and feet white. Flesh and skin white.

Weights

Male	4.10 kg (9 lb) min.
Female	3.20 kg (7 lb) min.

Scale of points

Type and flatness of back	25
Size	20
Legs and feet	15
Colour	20
Condition	10
Head and comb	10
	100

Serious defects (for which birds should be passed)

Other than four toes. Wry tail or any other deformity. Feather on shanks and toes. Rose comb.

BANTAM

Sussex bantams conform to the large fowl standard.

Weights

Male	1130 g (40 oz) max.
Female	790 g (28 oz) max.

Scale of points

Type, size and weight	35
Hackle, tail and wing	20
Body colour	15
Head and eye	15
Feet and legs	15
	100

TRANSYLVANIAN NAKED NECK

LARGE FOWL

Origin: Europe
Classification: Heavy: Rare
Egg colour: Tinted

The Translyvanian Naked Neck is stated to have been produced in what was formerly eastern Hungary, now largely included in Romania. The first examples were seen in Britain about 1880. They are regarded as one of the most vigorous of all breeds of poultry, the hens are excellent layers and good sitters and they are splendid foragers if allowed to roam, needing very little food during the greater part of the year. Naked Necks are a perfect example of the value of a rare breed as they are now one of the main breeds of the broiler industry, particularly in the hotter climate of France.

General characteristics: male

Carriage: Alert, upright and bold.
Type: Body large, deep and compact, well balanced and symmetrical. Back broad and of medium length, saddle hackle long and abundant. Breast broad and well rounded. Wings of medium size, carried well up. Tail medium size carried at an angle of 45°, sickles large and well curved.
Head: Medium size. Beak short, stout and slightly curved. Eyes large, bright and prominent. Comb single, medium in size, straight and erect, with well-formed spikes. Face smooth. Ear-lobes and wattles of medium size, fine in texture and smooth: the head to carry an oval cap of feathers surrounding the base of the comb, even in shape, with a small tassel at the back.
Neck: Of medium length, slightly curved, completely without feathering, stubs or fluff: the skin of the neck to be smooth and fine in texture, free from wrinkles or roughness. (A small tassel of feathers at the bottom of the neck above the breast feathers is permitted, but not desirable.)
Legs and feet: Legs of medium length, strong and stout. Shanks round and free of feathers. Toes, four, strong, straight and well spread.

Female

The general characteristics are similar to those of the male, allowing for the natural sexual differences.

Colour

The black

Male and female plumage: Dense black with a rich green sheen.

Black Transylvanian Naked Neck male, bantam

The white

Male and female plumage: Snow-white throughout.

The cuckoo

Male and female plumage: Colour as in Cochin.

The buff

Male and female plumage: Colour as in buff Rocks.

The red

Male and female plumage: Colour as in Rhode Island Reds.

The blue

Male and female plumage: Colour as in blue Minorca.

In both sexes and all colours

Eyes orange. Face, comb, ear-lobes, wattles and neck bright red. Legs, feet and beak yellow or horn in the black and cuckoo, yellow or white in the white.

Weights

Male	3.20–3.60 kg (7–8 lb)
Female	2.50–2.70 kg ($5\frac{1}{2}$–$6\frac{1}{2}$ lb)

Scale of points

Type and carriage	20
Head and neck	35
Legs and feet	10
Colour and markings	15
Size	10
Condition	10
	100

Serious defects

Any noticeable feather, fluff or stubs on the neck. Absence of cap of feathers on the head. Feathered legs. Other than four toes. Any deformity.

BANTAM

Bantam Naked Necks are to be replicas of their large fowl counterparts.

Weights

Male	910 g (32 oz)
Female	680 g (24 oz)

TUZO BANTAM

Origin: Japan
Classification: True bantam: Hard feather
Egg colour: Tinted

The Tuzo is a true hard feather bantam from Japan. It has been in this country since the early 1970s and is still in a few hands. It is not unlike a bantam Asil.

General characteristics: male

Carriage: Upright.
Type: Similar to a small Asil. Body broad at front with prominent shoulders. Wings short. Tail carried horizontally or a little below. Feathers short and hard.
Head: Broad and rounded with a short, slightly hooked beak. Well developed brows and protruding cheeks. Comb small triple or occasionally walnut. Wattles and lobes (if any) insignificant.
Neck: Strong and slightly curved.
Legs and feet: Thighs strong with slight bend at hock. Shanks medium strong and straight. Toes, four, fine and straight.

Female

The general characteristics are similar to those of the male, allowing for the natural sexual differences. The comb, if visible, should be small and insignificant.

Colour

Male and female plumage: Any Game colour.

In both sexes and all colours

Eyes light yellow to orange. Comb, face, wattles and lobes bright red.

Wheaten Tuzo female

Weights

Male	1020–1250 g (36–44 oz)
Female	790–1020 g (28–36 oz)

Scale of points

Type and carriage	40
Plumage quality	10
Colour	10
Head	20
Legs and feet	10
Condition	10
	100

Serious defects

Any deformity. Comb any other than standard.

VORWERK

LARGE FOWL

Origin: Germany
Classification: Light: Rare
Egg colour: Cream to tinted

Originated in Hamburg by Oskar Vorwerk in 1900, the breed was first shown at Hanover

Vorwerk female

in 1912 and standardized in 1913. The aim was to provide a middle-weight economical utility fowl, good natured, lively but not timid. A point worthy of note is the compatibility of males amongst themselves. These fowls were found to be particularly suitable for smallholdings and farmyards as they are excellent foragers, small eaters and quick maturing.

General characteristics: male

Carriage: Very powerful, compact utility shape, carriage low rather than high, not too much bone, markings the same in both sexes, lively but not timid.
Type: Body of considerable size, as broad and deep as possible like a rounded rectangle. Back broad, slightly sloping with a full saddle. Breast broad, deep and well rounded. Wings closely carried. Tall moderately tight, held at a lowish angle with well-rounded sickles of moderate length.
Head: Medium sized and moderately broad. Face covered with small feathers. Comb single, of medium size at the most, with four to six serrations. Wattles of medium length, well rounded. Lobes of barely average size. Eyes alert.
Neck: Of moderate length with full hackle and carried fairly upright, proudly.
Legs and feet: Moderate length with fine bone. Toes, four, small close fitting scales. Thighs fleshy and tightly feathered.
Plumage: Close fitting, glossy, velvety hackle.
Handling: Firm as befits an active forager.

Female

General characteristics are similar to those of the male allowing for the natural sexual differences. Back to be broad with almost no cushion. The latter part of the small comb may bend slightly to one side.

Colour

Male plumage: Head, hackle and tail should be velvety black. Body deep buff, under-colour grey. Wing secondaries buff; primaries dark grey to black. Saddle buff with light striping. Legs slate.
Female plumage: Hackle black with slight buff lacing permitted at the back of the head. Body and secondary wing flights buff; primaries greyish-black and buff mixed. Visible parts of the main tail black with the tail furnishings partly laced with buff. Undercolour grey.

In both sexes: Beak greyish-blue to horn. Eyes orange or orange-red. Comb, face and wattles red. Lobes white. Legs and feet slate.

Weights

Male	2.50–3.20 kg ($5\frac{1}{2}$–7 lb)
Female	2.00–2.50 kg ($4\frac{1}{2}$–$5\frac{1}{2}$ lb)

Scale of points

Type/utility quality	25
Head	10
Colour	25
Legs and feet	10
Size	15
Condition	15
	100

Serious defects

Body too narrow or too light. Carriage too high. Coarse bone. High tail. Lobes too red. Pale legs. In males, hackle unduly buff or grey, saddle nearly black. In females, lack of black in neck or tail and undue spangling in body feathers.

BANTAM

The same standard as in large fowl applies to bantams.

Weights

Male	910 g (32 oz)
Female	680 g (24 oz)

WELSUMMER

LARGE FOWL

Origin: Holland
Classification: Light: Soft feather
Egg colour: Brown to deep brown

Named after the village of Welsum, this Dutch breed has in its make-up such breeds as the partridge Cochin, partridge Wyandotte and partridge Leghorn, and still later the Barnevelder and the Rhode Island Red. In 1928, stock was imported into this country from Holland, in particular for its large brown egg, which remains its special feature, some products being mottled with brown spots. It has distinctive markings and colour, and comes into the light-breed category, although it has good body-size. It enters the medium class in the country of its origin. Judges and breeders work to a standard that values indications of productiveness, so that laying merits can be combined with beauty.

Welsummer male, large

Welsummer female, large

General characteristics: male

Carriage: Upright, alert and active.
Type: Body well built on good constitutional lines. Back broad and long. Breast full, well rounded and broad. Wings moderately long, carried closely to the sides. Tail fairly large and full, carried high, but not squirrel. Abdomen long, deep and wide.
Head: Symmetrical, well balanced, of fine quality without coarseness, excesses or exaggeration. Skull refined, especially at back. Beak strong, short and deep. Eyes keen in expression, bold, full, highly placed in skull and standing out prominently when viewed from front or back; pupils large and free from defective shape. Comb single, of medium size, firm, upright, free from any twists or excess around nostrils, clear of nostrils, and of fine, silky texture, five to seven broad and even serrations, the back following closely but not touching the line of the skull and neck. Face smooth, open and of silky texture, free from wrinkles or surfeit of flesh and without overhanging eyebrows. Ear-lobes small and almond-shaped. Wattles of medium size, fine and silky texture and close together.
Neck: Fairly long, slender at top but finishing with abundant hackle.
Legs and feet: Thighs to show clear of body without loss of breast. Shanks of medium length, medium bone and well set apart, free from feathers and with soft, pliable sinews, free from coarseness. Toes, four, long, straight and well spread out, back toe to follow in straight line, free from feathers between toes.
Plumage: Tight, silky and waxy, free from excess or coarseness, silky at abdomen and free from bagginess at thighs.
Handling: Compact, firm and neat bone throughout.

Female

The general characteristics are similar to those of the male, allowing for the natural sexual differences. Handling: Pelvic bones fine and pliable; abdomen pliable; flesh and skin of fine texture and free from coarseness; plumage sleek; abdomen capacious, but well supported by long breastbone and not drooping; general handling of a fit, keen and active layer.

Colour

Male plumage: Head and neck rich golden-brown. Hackles rich golden-brown as uniform as possible, free from black striping, yet underparts (out of sight) may show a little striping at present. Back, shoulder coverts and wing bow bright red-brown. Wing coverts black with green sheen forming a broad bar across (a little brown peppering at present permissible); primaries (out of sight when wing is closed), inner web black, outer web brown; secondaries, outer web brown, inner web black with brown peppering. Tail (main) black with a beetle-green sheen; coverts, upper black, lower black edged with brown. Breast black with red mottling. Abdominal and thigh fluff black and red mottled.
Female plumage: Head golden-brown. Hackle golden-brown or copper, the lower feathers with black striping and golden shaft. Breast rich chestnut-red going well down to the lower parts. Back and wing bow reddish-brown, each feather stippled or peppered with black specks (i.e. partridge marking), shaft of feather showing lighter and very distinct. Wing bar chestnut-brown; primaries, inner web black, outer brown; secondaries, outer web brown, coarsely stippled with black; inner web black, slightly peppered with brown. Abdomen and thighs brown with grey shading. Tail black, outer feathers pencilled with brown.

The silver duckwing

Male plumage: Head, neck and hackles white. Breast black with white mottling. Back shoulder coverts and wing bow white. Wing primaries flight feathers (out of sight as wings closed), inner web black, outer web white; secondaries, outer web white, inner web black, with white peppering, coverts black with green sheen forming a broad bar across pri-

maries. Tail, main black with beetle-green sheen; coverts, upper black, lower black, edged with white. Abdominal and thigh fluff black with white mottling.
Female plumage: Head and skull silvery white. Hackle silvery white and lower feathers with black striping, and white shaft. Breast salmon-red or robin-red. Back and wing bow silvery grey, each feather stippled or peppered with black specks (i.e. partridge marking), shaft of feather showing light and very distinct. Wing bar silvery grey; primaries, inner web black, outer web white; secondaries, outer web white, coarsely stippled with black, inner web black slightly peppered with white. Abdomen and thighs silvery grey. Tail black, outer feathers pencilled with white.

In both sexes and colours

Beak yellow or horn. Eyes red. Comb, face, ear-lobes and wattles bright red. Legs and feet yellow. Undercolour dark slate grey.

Weights

Cock	3.20 kg (7 lb);	cockerel	2.70 kg (6 lb)
Hen	2.70 kg (6 lb);	pullet	2.00–2.25 kg ($4\frac{1}{2}$–5 lb)

Scale of points

General type	20
Handling, size, and indications of productiveness	30
Head	10
Legs and feet	10
Colour	20
Condition	10
	100

Serious defects

Comb other than single or with side sprigs. White in lobe. Feather on legs, hocks or between toes. Other than four toes. Striping in neck hackle or saddle of male. Absolutely black or whole red breast in the male. Salmon breast in the female. Legs other than yellow. Badly crooked or duck toes. Any body deformity. Coarseness, beefiness and anything which interferes with the productiveness and general utility of the breed.

BANTAM

Welsummer bantams are to be miniatures of the large fowl and so the standard for large applies.

Weights

Male	1020 g (36 oz)
Female	790 g (28 oz)

WYANDOTTE

LARGE FOWL

Origin: America
Classification: Heavy: Soft feather
Egg colour: Tinted

The first variety of the Wyandotte family was the silver laced, originated in America, where it was standardized in 1883. The variety was introduced into England at the time, and our breeders immediately perfected the lacings and open ground colouring.

Partridge Cochin and gold spangled Hamburgh males were crossed with the silver females, to produce the gold laced variety. The white Wyandotte came as a sport from the silver laced; the buff followed by crossing buff Cochin with the silver laced. In 1896, the partridge variety was introduced from America, the result of blending partridge Cochin and Indian Game blood with that of the gold laced, the variety being perfected for markings in England. It was once called the gold pencilled, and the silver pencilled soon followed from partridge Wyandotte and dark Brahma crossings.

Columbians were the result of crossing the white Wyandotte with the barred Rock, and it was the crossing of the gold laced and the white varieties which produced the buff laced and the blue laced, first seen here in 1897. Blacks, blues and barred have been made in different ways in this country. The latest variety to be introduced is the red, created in Lancashire, from the gold laced variety, with selective matings with white Wyandotte, Barnevelder and Rhode Island Red.

It is clear that while the family of the Wyandotte is large, every variety is a made one from various blendings of breeds.

General characteristics: male

Carriage: Graceful, well balanced, alert and active, but docile.
Type: Body short and deep with well-rounded sides. Back broad and short with full and broad saddle rising with a concave sweep to the tail. Breast full, broad and round with a straight keel bone. Wings of medium size, nicely folded to the side. Tail medium size but full and spread at the base, the main feathers carried rather upright, the sickles of medium length.
Head: Short and broad. Beak stout and well curved. Eyes intelligent and prominent. Comb rose, firmly and evenly set on head, medium in height and width, low, and square at front, gradually tapering towards the back and terminating in a well defined spike (or leader) which should follow the curve of the neck without any upward tendency. The top should be oval and covered with small and rounded points; the side outline being convex to conform to the shape of the skull. Face smooth and fine in texture. Ear-lobes oblong, wattles medium length, fine in texture.
Neck: Of medium length and well arched with full hackle.
Legs and feet: Thighs of medium length, well covered with soft feathers; the fluff fairly close and silky. Shanks medium in length, strong, well rounded, good quality, and free of feather or fluff. Toes, four, straight and well spread.
Plumage: Fairly close and silky, not too abundant or fluffy.

Female

The general characteristics are similar to those of the male, allowing for the natural sexual differences.

Colour

The barred
Male and female plumage: Similar to that of the barred Plymouth Rock.

The black
Male and female plumage: Black with beetle-green sheen, undercolour as dark (black) as possible.

The blue
Male and female plumage: One even shade of blue, light to dark, but medium preferred; a clear solid blue, free from mealiness, 'pepper', sandiness, or bronze, and quite clear of lacing; a 'self colour' in fact.

The buff
Male and female plumage: Clear, sound buff throughout to skin, allowing greater

White Wyandotte female, bantam

Blue laced Wyandotte female, large

Partridge Wyandotte male, bantam

Partridge Wyandotte female, bantam

Silver pencilled Wyandotte female, bantam

Silver laced Wyandotte male, bantam

White Wyandotte male, bantam

lustre on the hackles and wing bow of the male. With these exceptions the colour should be perfectly uniform, but washiness or a red tinge, mealiness or 'pepper' to be avoided.

The columbian

Male and female plumage: Pearl-white with black markings. Primaries (wing) black or black edged with white; secondaries black inner web and white outer. The male's neck hackle broadly striped with intense black down the centre of each feather, such stripe entirely surrounded by a clearly defined white margin and finishing with a decided white point (free from black outer edging or black tips). Saddle hackle white. Tail glossy green-black, coverts either laced or not with white. The female's hackle bright intense black, each feather entirely surrounded by a well defined white margin. Tail feathers black, except the top pair which may or may not be laced with white. Remainder (in both sexes) white, entirely free of ticking, with slate or blue-white undercolour.

The partridge

Male plumage: Head dark orange. Hackles bright orange-yellow, shading to bright lemon-yellow, free from washiness, each feather having a clearly-defined glossy black stripe down the middle, not running out at the tip, and free from light shaft. Back and shoulders bright red of a scarlet shade, free from maroon or purple tint. Wing bar solid, glossy black; primaries solid black, free from white; secondaries, rich bay outer web and black inner and end of feather, the rich bay alone showing when the wing is closed. Undercolour black or dark grey, free from white. Breast and fluff metallic-black, free from red or grey ticking. Tail (including sickles and tail coverts) metallic-black, free from white at roots.

Female plumage: Head and hackle rich golden-yellow, the larger feathers finely and clearly pencilled. Breast, back, cushion and wings soft light partridge brown, quite even and free from red or yellow tinge, each feather plentifully and distinctly pencilled with black, the pencilling to follow the form of the feather, and to be even and uniform throughout. Fine, sharply defined pencilling with three or more distinct lines of black is preferred to coarse, broad marking, especially in females, in which the pencilling is generally better defined than in pullets. Pencilling that runs into the brown, peppery markings, and uneven, broken or barred pencilling, constitute defects. Light shafts to feathers on the breast must be penalized. Fluff brown (same shade as body), as clearly pencilled as possible. Primaries (wing) black; secondaries brown (same shade as body), pencilled with black on outer web, black on inner web, showing pencilling when the wing is closed. Tail black, with or without brown markings, with clearly-pencilled feathers up to the point of the tail.

The silver pencilled

Male and female plumage: Except that the ground is silver-white in the male and steel-grey in the female, instead of red, brown, etc. (of various shades), the silver pencilled is similar to the partridge.

The red

Male and female plumage: Surface rich, bright, glossy red. Neck hackle of medium shade to match body colour, with a black stripe down the centre of each feather at the lower part. Tail and coverts green-black. Wing primaries, inner half black, outer half red; secondaries inner half dark slate or black, outer half to match body. Undercolour dark or slate, clearly defined.

The white

Male and female plumage: Pure white, free from yellow or straw tinge.

In both sexes and all colours

Beak, legs and feet yellow or horn, which may dilute to straw in adults or laying pullets. Yellow preferred. Eyes bright bay, orange or red. Bright bay preferred.

The silver laced

Male plumage: Head silvery-white. Neck silvery-white with clear black stripe through the centre of each feather (white shaft is permissible), free from ticks. Saddle: hackles to match neck. Back silvery-white, free from yellow or straw colour. Shoulder tip white laced with black. Wing bow silvery-white; wing coverts evenly laced, forming at least two well-defined bars; secondaries black on inner and wide white strip on outer web, the edge laced with black; primaries or flights black on inner web and broadly laced white on outer edge. Breast and underparts: the web white with well-defined jet black lacing, free from double or white outer lacing, lacing regular from throat to back of thighs, showing green lustre. A shaft may be found in the laced feathers and is permissible but undesirable. Undercolour dark slate. Tail: true tail feathers, sickles and coverts black with green lustre. Thighs and fluff black slate with clear lacing round the hocks and outer side of thighs.
Female plumage: Head silvery-white. Neck silvery-white with clear black stripe through the centre of each feather (white shaft is permissible), free from ticks. Breast and back: undercolour dark slate, web white with regular, well-defined jet black lacing, free from double or outer lacing and showing green lustre. A shaft may be found in the laced feathers and is permissible but undesirable. Wings same as back on the broad portion; secondaries and primaries as in the cock. Tail black showing green lustre, the coverts black with a white centre to each feather. Thighs and fluff black or dark slate.

The other varieties are identical to the silver laced except for differences in ground colours and markings (i.e. lacings and stripings), and are as follows:

The gold laced

Ground colour rich golden bay; markings black.

The blue laced

Ground colour red-brown; markings clear blue.

The buff laced

Ground colour rich buff; markings white.

Note: In all colours and sexes, regularity of lacing to count above any breadth of lacing. Brightness and uniformity of ground colour to be considered of more value than any particular shade.

Scale of points for all laced colours

Lacing (including striping)	35
Colour	25
Head	10
Type	20
Condition	10
	100

Birds to be passed over for serious defects as in all colours, and points to be deducted for lack of presentation.

Weights

Cock not less than 4.08 kg (9 lb); cockerel not less than 3.62 kg (8 lb)
Hen not less than 3.17 kg (7 lb); pullet not less than 2.72 kg (6 lb)

Scale of points

The black

Colour (surface 25, undercolour 10)	35
Type	25
Head	10
Size and condition	15
Legs	15
	100

The blue

Type	25
Colour	25
Head	15
Legs	15
Size	10
Condition	10
	100

Serious defects: These include black legs devoid of yellow.

The buff

Type (back 10, body 12, wings 10, tail 8)	40
Head 6, comb 8, ear-lobes and wattles 8	22
Size and condition	20
Neck	10
Legs	8
	100

The columbian

Type or shape	25
Comb	10
Eye	5
Body colour	15
Hackle, including scantiness	10
Tail	5
Flights	5
Legs	5
Texture	5
Condition (including activity)	7
Size	8
	100

Serious defects (which should be heavily penalized): These include badly crooked breastbone, coarseness. Inactivity. Excess of feather. Overhanging eyebrows. Crooked toes. Brown undercolour. Green eyes.

The partridge and silver pencilled male

Colour and markings (colour of hackles 8, striping of hackles 8, top colour 8, breast 7, flights 5, tail 4, undercolour 4, fluff 4)	48
Head (comb 7, eyes 5, lobes and wattles 4)	16
Type	22
Size and condition	8
Legs	6
	100

The partridge and silver pencilled female

Colour and markings (ground colour 13, formation – breadth and form of black marks – of pencilling 11, clearness of pencilling 10, fluff 5, hackles 4)	43
Head (comb 6, eyes 5, lobes 4)	15
Type	22
Legs	10
Size and condition	10
	100

Serious defects: These include slipped wings, wall eyes or eyes that do not match.

The red

Indications of egg producing merits	15
Indications of reproductive merits	15
Type and carriage	20
Colour and markings	20
Head (including eyes)	10
Legs	10
Condition	10
	100

Serious defects: These include shanks other than yellow (allowance made for adult birds and heavy laying females). Coarseness, superfine bone. Any points against egg production, reproduction, or stamina values. Absence of any dark undercolour, and any deformity for which a bird may be passed.

The white

Type	25
Colour	25
Head	15
Legs and feet	10
Size	15
Condition	10
	100

Serious defects (for which a bird may be passed): These include feathers other than white in colour. Coarseness and 'Orpington' type.

Serious defects in all colours

Any feathers on shanks or toes. Permanent white or yellow in ear-lobe covering more than one-third of its surface. Comb other than rose, or falling over one side, or so large as to obstruct the sight. Shanks other than yellow, except in mature birds, which may shade to light straw. Any deformity.

BANTAM

Wyandotte bantams are miniatures of the large fowl and the standards in every respect are the same, with the exception of weights and some scales of points. White laced buffs, violet laced, blue laced, buff and cuckoo are also seen in bantams.

Colour

In both sexes: Beak bright yellow, except in marked and laced varieties, in which it may be horn, shaded with yellow. (*Note:* yellow beaks are unobtainable in black males with dark undercolour, and beak colour in these should be black, shaded with yellow.) Eyes bright bay in all colours. Comb, face, wattles and ear-lobes bright red. Legs and feet bright yellow.

Weights

Male	not to exceed 1.70 kg ($3\frac{3}{4}$ lb)
Female	not to exceed 1.36 kg (3 lb)

Scale of points

The white

Colour	25
Type	25
Head	5
Comb	10
Lobes	5
Eyes	5
Leg colour	5
Size	10
Condition	10
	100

The black

Colour	20
Undercolour	15
Type	20
Head	5
Comb and lobes	10
Legs colour	10
Size	12
Condition	8
	100

The columbian

Comb	10
Eyes	5
Lobes and wattles	5
Hackle and tail	15
Body colour	15
Legs	5
Type and symmetry	35
Condition	10
	100

Serious defects

Feathers on shanks or toes. Permanent white or yellow in ear-lobes, covering more than one-third of the surface. Comb other than rose or flopping or obstructing the sight. Shanks other than yellow. Any deformity. Slipped wings (which should be penalized strongly). Eyes not matching or other than bright bay. Conspicuous peppering on ground colour of laced varieties. Any form of double lacing in laced varieties.

YAMATO-GUNKEI

Origin: Japan
Classification: Hard feather: Rare
Egg colour: Cream or tinted

To complete the list of Japanese game introduced into this country comes the Yamato-Gunkei which is the largest of the small Japanese game. In fact, with its weight it could be regarded as 'intermediate'. It is an ancient ornamental game breed, popular around Tokyo. There were once several regional variants, but since World War II these have merged into a single overall type in Japan. The aim is to be as thick-set and exaggerated as possible while keeping to the weight limit. Plumage colour is not important; plumage is scanty. Mature birds have very wrinkled skin covering the face and throat.

General characteristics: male

Carriage: Upright, legs wide apart. Wings held away from the body at the shoulders.
Type: Back very broad and of medium length, basically flat but tapers down to the tail. The whole body very powerful and muscular. Breast broad. Tail short and carried low, pointing down between the legs and called 'prawn'-tailed in Japan
Head: Big, broad on top, with especially broad eyebrows. Beak hooked, short and strong. Wattles minimal, but ample dewlap showing bare skin well down the neck. Mature birds have wrinkled red skin covering the face. Eyes set in oval eyelids, partially obscured in old birds by the wrinked skin. Comb pea or walnut.
Neck: Medium length, very thick and muscular. Hackles only go about halfway down the neck.

Legs and feet: Legs of medium to shortish length, but not excessively short. Upper legs very muscular, showing prominently. Shanks very thick and rather square in cross-section. Four powerful toes, as straight as possible.
Plumage: Very hard and sparse. Bare red skin showing around the vent, at the wing joints on the back and especially down the keel. The shoulder coverts show clearly on the back giving the characteristic 'five hills'; this is seen across the back, looking from the head towards the tail, by one wing, shoulder coverts, the back, shoulder coverts, the other wing.

Female

Similar to the male, but seldom as exaggerated. The tail is short, straight, rather spread horizontally and carried well down.

Colour

The black
Glossy green-black throughout.

The black-mottle
Any mixture of black and white. Called 'Go-stone' in Japan after the bowl of black and white counters used in the popular game of 'Go'.

The black-red
The red in cocks may be any shade from yellow to dark red. Females may be wheaten or cinnamon of any shade.

The blue
May be laced or self.

The buff
Normally a light shade called 'orang-utan' in Japan.

The spangled
Any mixture of black, white and shades of red.

The duckwings and greys
Normally with silver wheaten females.

The cuckoo, the white
As in Game.

In both sexes and all colours
Eyes pearl or yellow. Beak yellow or horn. Legs yellow or dusky in darker varieties. Head, skin, keel, etc. red.

Weights

Cock	2.0 kg ($4\frac{1}{2}$ lb);	cockerel	1.5 kg ($3\frac{1}{2}$ lb)
Hen	1.7 kg (4 lb)	pullet	1.3 kg (3 lb)

Scale of points

Carriage and type	30
Head and neck	20
Condition and plumage quality	20
Legs and feet	10
Eye colour	10
Legs and feet colour	5
Plumage colour	5
	100

YOKOHAMA

LARGE FOWL

Origin: Japan
Classification: Light: Rare
Egg colour: Tinted

The earliest long-tailed fowls were found in China. Then during the seventeenth century more were found in Japan. These were in various colours, similar to Sumatras with walnut and pea combs, the name given to these was Satumadori. From these the very long-tailed Onagadori evolved, and the traditional custom was to collect the tail feathers for decoration. The breed now receives government support to the top breeders and is recognized as a living national heritage. The Red-saddled variety was developed in Brunswick, Germany by Herr Hugo de Roi in the late-nineteenth century and is not standardized in the Far East. The Japanese are perfectionists in keeping the tails growing, which is at least one metre per season, the breed requiring the correct lighting and heating to maintain the growth throughout the year. In this country under natural conditions the bird will moult in the normal way.

General characteristics: male

Carriage: Stylish and pheasant-like.
Type: Body fairly long and deep, full round breast, long back tapering to tail, long wings carried rather low but close to the sides. Tail as long and flowing as possible, with a great abundance of side hangers, the sickles and coverts narrow and the whole tail forming a graceful curve and carried low.
Head: Skull small but inclined to be long and tapering. Beak strong and curved. Eyes bright and full of life. Comb single, pea or walnut, small and even. Face of fine texture. Ear-lobes small, oval and almond shape, fitting closely. Wattles round and small.
Neck: Long and furnished with flowing hackle, forming a complete circle around the neck.
Legs and feet: Of medium length, the shanks fine and free of feathers. Toes four, well spread.

Female

The back to be long, tapering to the tail and furnished with long saddle hackles. The tail must be very long and carried horizontally with the two top feathers gracefully curved and the coverts sickle-like. The remaining general characteristics are similar to those of the male, allowing for the natural sexual differences.

Colour

The golden or black-red

Male plumage: Head bay, hackles red, each feather having a black stripe in the centre. Breast, thighs, tail and fluff black, with beetle-green lustre. Back red. Wing bow and secondaries deep bay, bar black.
Female plumage: Head brown, hackle orange, each feather striped with black in the centre. Breast salmon, a lighter shade below. Thighs grey-brown. Back and shoulders brown, each feather with a light shaft. Tail coverts brown. Tail black, the two uppermost main feathers spotted or grizzled with light brown.

The duckwings (silver and gold)

Male and female plumage: As in the corresponding colours in Game fowl.

In both sexes: Beak horn. Eyes ruby-red. Comb, face and wattles bright red. Ear-lobes white or red. Legs and feet yellow, willow or slate-blue.

Red saddled Yokohama male, large

Black-red Yokohama male, bantam

White Yokohama male, bantam

The red saddled

Male and female plumage: White and red (practically a pyle). Breast and thighs red in the male and red-buff in the female, with distinct white spangling at the end of each feather. Male's back and wing bow crimson-red, the former vignetted into the saddle. Remainder white.

In both sexes: Beak yellow. Eyes ruby-red. Comb, face, wattles and ear-lobes bright red. Legs and feet bright yellow.

The white

Male and female plumage: Snow-white, free of any straw tinge.

In both sexes: Beak, legs, and feet white or yellow. Eyes bright red. Comb, face, wattles and ear-lobes red.

Other colours: The foregoing are the principal colours, and others are not kept or bred in sufficient numbers to warrant description. The above colours and markings are ideal, but general type, quality and length of the tail and hackles are the most important points in the Yokohama fowl.

Weights

Male 1.80–2.70 kg (4–6 lb)
Female 1.10–1.80 kg ($2\frac{1}{2}$–4 lb)

Scale of points

Quality and length of tail and number of feathers	25
Quality and length of neck and saddle hackles	20
Type and carriage	20
Head	10
Colour	5
Condition	5
Legs and feet	5
Size	10
	100

Serious defects

Yellow or straw coloured feathers in the white. White in face. Broken tail feathers. Short saddle hackles. Too high tail carriage. Any deformity.

BANTAM

Bantam Yokohamas should follow exactly the large fowl standard.

Weights

Male	570–680 g (20–24 oz)
Female	490–570 g (16–20 oz)

OTHER BREEDS

Belgian game

There are two varieties, the Bruges and the Liège. The breed was already known in the Middle Ages as the Bruges Blue after its blue plumage and the Flemish city. The main features of the Bruges are a straight horizontal back, upright stance, single comb and dark legs, eyes and beak. The Liège was produced in the nineteenth century due to rivalry between the Flemish and Walloon people, and the Walloon cockers crossed the Bruges with Malay to give a gamefowl with a long reach, and of 12 lb plus in weight. A third gamefowl, the Tienen, is similar to the Liège but has white legs and beak, fiery red eyes and cuckoo plumage. Bantams of the three varieties were known in 1943.

Brabanter

A Dutch breed, classified as light, which has a forward pointing crest and a cup comb. The colours are similar to the Appenzeller Spitzhauben.

Bresse

A French breed which is famous for its table qualities, but possibly because of its dark shanks, has not become popular in the UK.

Italiener

This is the German form of the Leghorn.

Rhienlander

A German breed which is rose combed and has white ear-lobes and long wattles. It comes in black, white, blue and cuckoo in both bantam and large. It has a long body, a long tail, broad feathering and shortish legs.

Uilebaard

A Dutch breed with a horn comb, a beard and a tassel. Its name translated is Dutch Owl Beard and it is a light breed.

Turkeys

Turkeys were introduced to Europe in the earlier part of the sixteenth century following their discovery by Spanish explorers of the Americas. Classes for turkeys were offered at the first English Poultry Show in 1845, and a standard for turkeys appeared in the first English Book of Standards in 1865. The standards published here relate solely to the traditional type pedigree breeds of turkey, which are naturally bred and not related in any way to the modern day, commercial, broad breasted or dimple breasted types, which are not considered appropriate in the show-pen.

GENERAL STANDARD: HEAVY BREEDS

General characteristics: male and female

Carriage: Stately and moderately upright.
Type: Body long, deep and well rounded. Back curving with good slope to tail. Breast broad, full, long and straight. Wings strong and large. Tail long in proportion to body.
Head: Long, broad and carunculated. Beak strong, curved and well set. Eyes bright, bold and clear. Throat wattle large and pendant.
Neck: Long, curving backward towards tail.
Legs and feet: Thighs long and stout. Fluff short. Shanks large, strong, well rounded and of medium length. Toes, four, straight and strong and well spread.

GENERAL STANDARD: LIGHT BREEDS

The general characteristics are the same as for the Heavy breeds with the following exceptions.
Carriage: Active and upright.
Legs and feet: Shanks large, strong and fairly long.

BOURBON RED

Classification: Light

Male plumage: The head and neck are brownish-red. Flights solid white. Tail feathers white with indistinct red bar. Neck brownish-red. Remainder of plumage rich, dark, brownish-red, each feather narrowly edged with black.
Female plumage: Similar to the male but with no black edgings. Narrow white edging on the breast.

In both sexes: The eyes are hazel brown, beak, legs and feet horn. Face, jaws, wattle and caruncles bright rich red.

Weights:

Stag 9.97–12.70 kg (22–28 lb); young stag 7.25–10.43 kg (16–23 lb)
Hen 5.44–8.16 kg (12–18 lb); young hen 3.62–6.35 kg (8–14 lb)

BRITISH WHITE

Classification: Heavy

Plumage of both sexes: Pure white with black tassel. Beak white to pale horn. Eyes, iris dark hazel, pupil blue-black. Face, wattle and caruncles, bright rich red, but changeable in the male to blue and white. Shanks and feet pink flesh. Toenails white to pale horn.

Weights

Stag 12.70 kg (28 lb) upward, but not to exceed 17.23 kg (38 lb);
young stag 8.16–12.70 kg (18–28 lb)
Hen 7.25–9.97 kg (16–22 lb); young hen 6.35–8.16 kg (14–18 lb)

BRONZE

Classification: Heavy

Plumage of both sexes: A good metallic-bronze throughout, the female to have a slight ticking on the breast. Flights black with a definite white barring. Tail black and brown, with a good broad black band edged with white. Beak horn. Eyes, iris a dark hazel, pupil blue-black. Face, jaws, wattle and caruncles bright rich red. Shanks and toes black or horn. Toenails horn.

Weights

Stag 13.60–18.14 kg (30–40 lb); young stag 11.33–15.87 kg (25–35 lb)
Hen 8.16–11.79 kg (18–26 lb); young hen 6.35–9.97 kg (14–22 lb)

BUFF

Classification: Light

Plumage of both sexes: A deep cinnamon-brown. Flights and secondaries white. Tail deep cinnamon-brown edged with white. Beak light horn. Eyes, iris dark hazel, pupil blue-black. Face, jaws, wattle and caruncles bright rich red. Shanks and toes pink and flesh. Toenails light horn.

Weights

Stag 9.97–12.70 kg (22–28 lb); young stag 7.25–10.43 kg (16–23 lb)
Hen 5.44–8.16 kg (12–18 lb); young hen 3.62–6.35 kg (8–14 lb)

CAMBRIDGE BRONZE

Classification: Heavy

Plumage of both sexes: A dull bronze, with grey and white tips to the body feathers, and bars on the flights and tail as in the Bronze. Eyes dark brown. Beak flesh to horn. Legs and feet dark grey.

Weights

Stag 8.16–10.88 kg (18–24 lb)
Hen 5.44–7.25 kg (12–16 lb)

Bronze Turkey, male

Buff Turkeys, male and female

CRIMSON DAWN OR BLACK WINGED BRONZE

Classification: Heavy

Plumage of both sexes: As in Bronze, except that the primary flight feathers are black, the secondary flights black with white tips, and some white marking is permitted on the shoulders. All other aspects as in Bronze.

Weights

Stag	13.60–18.14 kg (30–40 lb);	young stag	11.33–15.87 kg (25–35 lb)
Hen	8.16–11.97 kg (18–26 lb);	young hen	4.98–9.97 kg (11–22 lb)

CRÖLLWITZER

Classification: Light

Plumage in both sexes: Head and neck white. Breast and back white with every feather ending in a black edge with a fine line of white outside. Tail and tail coverts white with a black band near the end of each feather. Flights white with black edge. Eyes hazel brown. Beak, legs and feet flesh to horn.

Weights

Stag 7.25–12.70 kg (16–28 lb)
Hen 3.62–8.16 kg (8–18 lb)

NORFOLK BLACK

Classification: Light

The original turkey importations into this country were darks and, no doubt, the first variety to be developed here was the Norfolk Black in the county from which it derived its name. Specimens were shipped to America and the breed may well be claimed as the first of the domestic varieties of turkey.

General characteristics: male and female

Type: Body fairly long and deep, particularly broad across the shoulders. Back broad and flat between the shoulders. Breast not too long, well rounded, muscular and fleshy. Wings carried lightly. Tail long in proportion to body.
Head: Fairly long and broad, and carunculated. Beak strong curved and well set. Eyes bright, bold and clear. Throat wattle large and pendant.
Neck: Of medium length curving slightly backward with an alert carriage.
Legs and feet: Legs short to medium length and well set apart. Thighs full and thick. Toes, four, straight, strong and well spread.

Colour

Plumage of both sexes: A dense black.

In both sexes: Beak, legs and feet black. Eyes dark to black. Face, jaws, wattle and caruncles bright rich red; but short black feathers on head and face not a fault.

Weights

Stag	11.35 kg (25 lb);	young stag	8.15–10.00 kg (18–22 lb)
Hen	5.90–6.80 kg (13–15 lb);	young hen	5.00–5.90 kg (11–13 lb)

Cröllwitzer Turkey, male

Norfolk Black Turkey, male

Slate Turkey, male

SLATE OR BLUE

Classification: Light

Colour

Plumage of both sexes: Dark or light shade of sound and even blue, free from brown feathers. Beak, legs and feet slate-blue. Eyes dark to black. Face, jaws, wattle and caruncles bright rich red.

Weights

Stag 8.16–11.33 kg (18–25 lb)
Hen 6.35–8.16 kg (14–18 lb)

ALL BREEDS

Scale of points

Type, carriage and size	40
Head	20
Legs and feet	10
Colour	20
Condition	10
	100

Serious defects

In all breeds, crooked breast, wry tail or any other deformity. Pronounced debeaking. Dimple breasts. In blues, presence of black or brown feathers. In Bourbon Reds pale ground colour, any black ticking or flecking. In British Whites any coloured feathers. In Buffs white in tail except in edging. In all Bronze varieties any wholly white feathers. In Norfolk Blacks any bronze in undercolour or wings, white flecks on thigh or wing feathers.

OTHER BREEDS

Among other breeds of turkey occasionally exhibited in this country may be mentioned the Lavender, the American Slate, the Beltsville White, the Narragansett, the Pied, the Nebraskan Spotted, the Cornish Palm and the Royal Palm.

Geese

The first poultry show of 1845 classified Common Geese, Asiatic or Knob Geese and 'Any other variety'. The first Book of Standards described the Toulouse and Embden. Peculiarly enough, these two breeds monopolized our standards up to recent times, being the chief ones exhibited regularly at shows. At times, other breeds have been exhibited and now the standards have been extended. The Greylag is said to be the ancestor of all our domestic geese, and the common goose of this country was undoubtedly the English Grey, although a white variety existed, and the Grey Back (Saddleback) may have come from an intercross.

The British Waterfowl Association classifies the following as wild geese: Canada, Egyptian and all species of British or foreign wild geese.

AFRICAN

Origin: China

The African goose is the large relative of the Chinese, both having been developed from the wild swan goose which, of course, they both resemble.

General characteristics: male and female

Carriage: Reasonably upright.
Type: Body large, long, carried moderately upright. Nearly the same thickness from back to front. Ideally, the underline is smooth and free from keel. Lobes even. Stern round, full, free from bagginess. Back broad, moderately long, flat. Breast full, well rounded, carried moderately upright, without keel. Wings large, strong, smoothly folded against sides. Tail well elevated.
Head: Broad, deep, large. Dewlap large, heavy, smooth; lower edges regularly curved and extending from lower mandible to below juncture of neck and throat. Bill rather large, stout at base. Knob large, broad as the head, protruding slightly forward from front of skull at upper mandible. Eyes large.
Neck: Long, nicely arched; throat with well-developed dewlap.
Legs and feet: Lower thighs short, stout. Shanks of medium length. Straight toes connected by web.

Colour

Plumage of both sexes: Head light brown. Neck very light ashy brown with distinct broad, dark brown stripe down centre of the back of the neck and extending its entire length. Front of neck under mandible very light ashy brown, gradually getting lighter in colour until past the dewlap where it is almost cream in colour, then gradually deepening in colour as it approaches the breast. Breast very light ashy brown shading to a lighter colour under the body. Body a lighter shade than the breast, gradually getting lighter as it approaches the fluff, which is so light as to approach white; sides of body ashy brown, each feather edged with a lighter shade. Lower thighs: upper part similar to sides of body, ashy brown edged with a lighter shade, lower part similar in colour to underpart of body. Back ashy brown. Wing bow ashy brown, slightly edged with a lighter shade; coverts ashy brown, distinctly edged with a lighter shade; primaries dark slate; primary coverts light slate; secondaries dark slate, distinctly edged with a lighter shade approaching white. Tail ashy brown heavily edged with a shade approaching white, tail coverts white.

In both sexes: Bill and knob black. Eyes dark brown. Legs and webs dark orange.

Weights

Gander 9.97–12.70 kg (22–28 lb)
Goose 8.16–10.88 kg (18–24 lb)

Scale of points

Size	20
Colour	10
Condition	10
Legs and feet	5
Breast	10
Head and throat	15
Neck	5
General carriage	15
Tail and paunch	10
	100

Defects

Lack of dewlap; lack of knob; white in coloured plumage.

WHITE AFRICAN

White African geese are identical to grey African geese in shape requirements, but the colour details are similar to those specified for white Chinese.

AMERICAN BUFF

Origin: America

The American buff goose is a heavy breed with a stance like the Embden, smooth breasted and dual lobed.

General characteristics: male and female

Carriage: Upright.
Type: Body moderately long, broad, plump. Back medium length, broad and smooth. Breast broad, deep and full. Wings medium in size and smoothly folded close to body. Tail medium in length with broad, stiff feathers.
Head: Broad, oval, strong. Bill of medium length, stout and tapering evenly to well-rounded end. Eyes large and full.
Neck: Rather upright and strong in appearance.
Legs and feet: Lower thighs medium length, well fleshed. Shanks stout, straight, moderately long. Toes straight and well webbed.

Colour

Plumage of both sexes: A rich shade of buff throughout with markings similar to the Toulouse.

African male

American Buff male

Bill orange. Eyes dark hazel - legs and webs orange.

Weights

Gander 9.97–12.70 kg (22–28 lb)
Goose 9.07–11.79 kg (20–26 lb)

Scale of points

Size	15
Colour	25
Condition	10
Legs and feet	5
Breast	10
Head and throat	10
Neck	5
Carriage	10
Tail and paunch	10
	100

Defects

White feathers in coloured plumage excepting white around the bill with age. Uneven lobes back and front.

Brecon Buff female

BRECON BUFF

Origin: Great Britain

At different times attempts have been made to create a buff goose, and to Wales goes the credit of originating the Brecon Buff, founded on stock from Breconshire hill farms, the breed being recognized officially in 1934. Hard, light in bone, with maximum flesh, and an active breed, it is also more prolific than the ultra-heavy breeds.

General characteristics: male and female

Carriage: Upright and alert, indicative of activity.
Type: Body broad, well rounded and compact. Breast round. Dual-lobed paunch. Wings strong.
Head: Neat, no sign of coarseness. Eyes bright.
Neck: Medium, the throat showing no gullet.
Legs and feet: Legs fairly short. Strong shanks. Straight toes connected by web.
Plumage: Hard and tight.

Colour

Plumage in both sexes: A deep shade of buff throughout with markings similar to those of the Toulouse. Ganders are usually not as deep coloured as geese.

In both sexes: Bill pink. Eyes dark brown. Legs and webs pink.

Weights

Gander 7.25–9.07 kg (16–20 lb)
Goose 6.35–8.16 kg (14–18 lb)

Scale of points

Size	10
Colour	25
Condition	10
Head and throat	10
Neck	5
Breast	15
Tail and paunch	10
Legs and feet	5
Carriage	10
	100

Defects

Uneven lobes.

BUFF BACK

Origin: Europe

General characteristics: male and female

Carriage: Nearly horizontal.
Type: Body moderately long, plump, deep and meaty with no evidence of a keel. Back slightly convex and approximately 60% broader than deep. Paunch moderately broad, deep and double-lobed. Wings rather long with the tips crossing over the tail coverts. Carried high and smoothly folded. Tail somewhat short, closely folded and carried nearly level.
Head: Fairly broad and somewhat refined with a nearly flat crown. Bill medium in length, nearly straight and stout. Eyes large and rather prominent.
Neck: Medium length, moderately stout and carried rather upright with little or no indication of an arch.
Legs and feet: Lower thighs medium length, plump and nearly concealed by ample thigh coverts. Shanks moderately long and rather refined, yet sturdy.

Colour

Plumage of both sexes: Head buff. Neck buff on the upper part and white on the lower. Back buff (in a heart shape) with each feather edged with near white. Breast, body, wings and tail white, except in some specimens for a broad band of buff edged with near-white beginning under the secondaries (just over the shanks) and extending under the abdomen to the opposite wing. Remainder of plumage white, except for large thigh coverts, which are buff and edged with white.

In both sexes: Bill orange. Eyes blue. Legs and webs orange.

Weights

Gander	8.16–9.97 kg (18–22 lb)
Goose	7.25–9.07 kg (16–20 lb)

Scale of points

Size	10
Colour and markings	35
Condition	10
Legs and feet	5
Breast	10
Head and throat	5
Neck	5
General carriage	10
Tail and paunch	10
	100

CHINESE

Origin: Asia

A prolific breed of the smaller-bodied type of goose.

General characteristics: male and female

Carriage: Upright, compact and active.
Type: Body compact and plump. Back reasonably short, broad, flat, and sloping, to give a characteristic upright carriage. Breast well rounded and plump, and carried high. Wings large, strong and high-up, carried closely. Stern well rounded, and well-developed paunch. Tail close and carried well out.
Head: Medium for size, and proportionate. Bill stout at base, symmetrical, medium for size. Knob, large rounded and prominent (smaller in the goose than the gander). Eyes bold.
Neck: Long, carried upright, but with graceful arch and refined (longer in the gander than the goose).
Legs and feet: Legs medium. Shanks strong, medium for length. Toes straight, well spread and webbed.
Plumage: This should be reasonably tight.

Colour

The white

Plumage of both sexes: Pure white.
In both sexes: Bill and knob orange. Eyes blue. Legs and webs orange-yellow.

The brown-grey

Plumage of both sexes: Head dark brown, with face fawn up to demarcation line above the eyes. Face band a definite white band or line from top of head to as far down the face as possible. Neck fawn with prominent dark brown stripe down middle of back of neck and for its entire length. Back brown. Breast greyish-fawn from under mandible well down to body where it becomes lighter. Thighs at side of breast brown, each feather edged with a lighter shade of greyish-fawn, approaching white. Wing bow and coverts medium brown, each feather laced with a lighter greyish-fawn edging, approaching white; flights brown. Stern, paunch and tail a lighter shade of greyish-fawn, approaching white, the tail having a broad band of russet-brown across with the light edging.
In both sexes: Bill and knob black. Eyes brown. Legs and web dull orange.

Weights

Gander	4.55–5.45 kg (10–12 lb)
Goose	3.60–4.55 kg (8–10 lb)

White Chinese male

Grey Chinese male

Scale of points

Size	10
Colour	15
Condition	10
Head and throat	20
Neck	10
Breast	5
Tail and paunch	5
Legs and feet	5
Carriage	20
	100

Defects

Absence of knob; any sign of a gullet. White in coloured plumage.

EMBDEN

Origin: North Europe

As the Embden breed was also known originally as the Bremen, one associates it with Germany although stock reached us from Holland. In Germany and North Holland, no doubt they crossed the Italian white with their native whites, creating the Embden. When stock did reach this country our breeders crossed the birds with our own English whites,

Embden male

and by careful selection increased the body weight and quantity of meat, while standardizing the breed for characteristics.

General characteristics: male and female

Carriage: Upright and defiant.
Type: Body broad, thick and well rounded. Back long and straight. Breast round. Shoulders and stern broad. Paunch deep and dual lobed. Wings large and strong. Tail close and carried well out.
Head: Strong, bold. Stout bill. Eyes bold.
Neck: Long, well proportioned, without a gullet.
Legs and feet: Legs medium. Shanks large and strong. Toes straight and connected by web.
Plumage: Hard and tight.

Colour

Plumage of both sexes: Pure glossy white.
In both sexes: Bill orange. Eyes light blue. Legs and webs bright orange.

Weights

Gander 12.70–15.42 kg (28–34 lb)
Goose 10.88–12.70 kg (24–28 lb)

Scale of points

Size	20
Colour	10
Condition	10
Head and throat	12
Neck	10
Breast	10
Tail and paunch	10
Legs and feet	6
Carriage	12
	100

Defects

Plumage other than white. Uneven lobes; indication of a keel. Any deformity.

GREY BACK

Origin: Germany

This is similar in type to the Buff Back. Unlike the grey-back Pomeranian it is dual lobed in the paunch.

PILGRIM

Origin: Great Britain. First standardized in the USA

General characteristics: male and female

Carriage: Above the horizontal, but not upright.
Type: Body moderately long, plump and meaty; keel permissible in goose. Adult abdomen deep, square and well balanced, free from bagginess. Back moderately broad, uniform in width, flat and straight. Breast round, full, deep. Wings strong, well developed, neatly carried to body. Tail medium in length, closely folded, carried nearly level.
Head: Medium in size, oval, trim. Bill medium in length, straight, stout, smoothly attached. Eyes moderately large.
Neck: Medium in length, moderately stout, slightly arched.
Legs and feet: Lower thighs medium in length, well fleshed. Shanks moderately short and stout. Toes, strong, straight, and well webbed.
Plumage: Hard, tight and glossy.

Colour

Gander's plumage: Pure white. Some hidden grey permissible in back plumage, and in wings, back and tail of young ganders.
Goose's plumage: Head light grey, the forepart broken with white, the white becoming more extensive with age. All white heads objectionable at any age. Neck light grey, upper portion mixed with white in mature specimens. Back light ashy grey, laced with lighter grey. Breast very light ashy grey, gradually getting lighter as it approaches fluff which is so light as to approach white. Sides of body soft, ashy grey, each feather edged with lighter shade. Wing bow and coverts light ashy grey, edged with lighter grey; primaries medium

Pilgrim male and female

grey; coverts light grey; secondaries medium grey with lighter shade approaching white. Tail ashy grey, heavily edged with a lighter grey approaching white.

In both sexes: Bill orange. Eyes bluish-grey in gander, hazel-brown in goose. Legs and webs orange.

Weights

Gander	6.35–8.16 kg (14–18 lb)
Goose	5.44–7.25 kg (12–16 lb)

Scale of points

Size	10
Colour	35
Condition	10
Head and throat	10
Neck	5
Breast	10
Tail and paunch	5
Legs and feet	5
Carriage	10
	100

Defects

Flesh- or pink-coloured bills, feet and shanks. Single-lobed or unbalanced paunches. Solid features in the plumage of the gander. White flights, all white heads and white blaze (on the breast) in females. Predominantly white necks to be subject to severe discrimination. Undersize in either sex.

Pomeranian male

POMERANIAN

Origin: Europe

General characteristics: male and female

Carriage: Nearly horizontal.
Type: Body moderately long, plump, deep and meaty. Keel not permissible. Paunch moderately deep, broad, single lobed. Breast plump and broad. Wings long with tips crossing over tail coverts, carried high neatly and smoothly folded. Tail short, closely folded, carried nearly level. Back slightly convex. Two-thirds more length than breadth.
Head: Fairly broad, refined, crown somewhat flat. Bill medium length stout and nearly straight. Eyes large and prominent.
Neck: Medium in length, stout and carried upright.
Legs and feet: Lower thighs medium length, plump, nearly covered by ample thigh coverts. Shanks moderately long. Feet and toes straight and well webbed.

Colour

Plumage of both sexes: Solid coloured heads preferred but some specimens have white feathers around the base of the bill. Neck upper dark grey, lower white. Back dark grey edged with grey-white from a point above the scapulars to near the base of the tail. Scapulars the same colour. This should suggest a heart shape. Tail, wings, breast and body white except for a broad band of dark grey edged with near white beginning under the secondaries, just over the shanks and extending under the abdomen to the opposite wing (the latter is only found in some specimens).

In both sexes: Bill reddish-pink or deep flesh colour. Shanks and feet orange-red. Eyes blue.

On the Continent there is also Buff and Solid Grey and Solid White.

Weights

Gander 8.16–10.88 kg (18–24 lb)
Goose 7.25–9.07 kg (16–20 lb)

Scale of points

Size	10
Colour	35
Condition	10
Head and throat	5
Legs and feet	5
Tail and paunch	10
Breast	10
Carriage	10
Neck	5
	100

Defects

Dual-lobed paunch.

ROMAN

Origin: Mediterranean

Another of the smaller type of goose, the Roman was introduced into England from Italy about 1903, and there were other importations at later dates. Earliest arrivals often were marked with grey on the back, but were eliminated by selective matings for the pure white. (Breed illustration on page 304.)

General characteristics: male and female

Carriage: Active, alert, with horizontal outline.
Type: Compact and plump, deep and broad, and well balanced. Back wide and flat. Breast full, well rounded, somewhat low and without keel. Wings long, strong, high-up and well tucked up to tail line. Stern well rounded off, paunch not too pronounced, dual lobed. Tail close, long and carried well out.
Head: Neat and well rounded symmetrical and refined. Face deep. Bill short and not coarse. Eyes bold, well up in skull.
Neck: Upright, medium length, refined (particularly in goose) and without gullet.
Legs and feet: Legs short, light boned, well apart. Toes straight and connected by web.
Plumage: Sleek, short, tight and with glossy feathering.

Colour

Plumage of both sexes: Glossy white.
In both sexes: Bill orange-pink. Eyes light blue. Legs and webs orange-pink.

Weights

Gander 5.45–6.35 kg (12–14 lb)
Goose 4.55–5.45 kg (10–12 lb)

Scale of points

Size	15
Colour	10
Condition	10
Head and throat	10
Neck	10
Breast	15
Tail and paunch	10
Legs and feet	5
Carriage	15
	100

Defects

Plumage other than white. Any deformity. Excessive weight or bone. Coarseness and oversize.

SEBASTOPOL

Origin: Eastern Europe

The Sebastopol is one of the most unusual of the breeds of domestic geese. The long frizzled or spiralled feathers on the breast and the loose fluffed plumage make the Sebastopol a unique and attractive breed.

The breed is primarily one for the exhibitor but the Sebastopol is a moderate egg-layer and a fast grower and thus has merit as a utility goose.

The Frizzle

Carriage: Horizontal.
Type: Body appears round because of the full feathering. Back of medium length but appears short because the long feathers give the body a rounded ball appearance. Breast full and deep, lacking keel. Wing feathers long, well curled and flexible. They make the bird incapable of flight because they are devoid of stiff shafts. Tail composed of long, well-curled feathers.
Head: Neat head. Bill of medium length. Eyes large and prominent.
Neck: Medium length and carried rather upright instead of forward.
Legs and feet: Lower thighs short but stout, each covered with curled feathers. Shanks short and stout.
Plumage: Only feathers of head and upper neck smooth. Feathers on lower neck, breast and remainder of body profusely curled. Feathers of wings and back should be long (the longer the better), well curled and free from stiff shafts. Specimens of good stock and in good condition should display back and wing feathers that should almost touch the ground.

Colour

The white
Plumage of both sexes: Pure white in colour, though traces of grey in young females allowed.

The buff
Plumage of both sexes: Even buff colour interrupted by curled feathers.

The Smooth

Carriage: Horizontal.

Roman male

Sebastopol female

Type: Body comparatively short. Back wide and rounded, sloping gently from shoulders to the tail. Breast smooth, full and well rounded, without any keel. Paunch neat and smooth, dual lobed; not heavy or sagging. Tail short and held closed, carried horizontally or a little elevated.
Head: As the Frizzle.
Neck: As the Frizzle.
Legs and feet: Thighs short and strong. Shanks short and stout. Toes straight and connected by web.
Plumage: The feathers of the head, neck, breast, belly and paunch smooth. Feathers of the back, saddle, shoulders, wing bow and wing coverts are broad and extended in length; profuse, loosely curled and spiralled, falling over the wings and rump, often trailing along the ground. Feathers covering the wing fronts and thighs broad and extended, loosely curled and falling to the ground. The shafts of the primaries and secondaries are soft and flexible; the barbs of these feathers are 'fluted' giving a slight wave to the edge of vane. The tail is made up of unevenly set 'fluted' feathers. The plumage to obscure the legs and feet from view.

Colour

As in the Frizzle.

In both colours and types

Bill and webs orange. Eyes bright blue.

Weights

Gander	5.44–7.25 kg (12–16 lb)
Goose	4.53–6.35 kg (10–14 lb)

Scale of points

Size	10
Colour	10
Conditioning and feathering	40
Head and throat	5
Neck	5
Breast	5
Tail and paunch	5
Legs and feet	5
Carriage	15
	100

Defects

Silky feathering on the back of the Frizzle. Angel wings in either type.

STEINBACHER

Origin: Area of Thuringen (former East Germany)

Bred from a cross of local and Chinese geese at the end of the last century. Since 1932 acknowledged as a breed.

General characteristics: male and female

Carriage: Medium-sized goose, strong, with proud stature, straight neck.
Type: Gander: Back strong, slightly stocky. Back, wide sloping downwards towards the

Steinbacher male

back. Tail, short, pointed and carried level. Breast, wide and full, not too high. Belly and rear part, full and wide, well developed. If possible without paunch in young ganders, for old ganders not too heavily marked paunch is acceptable. Wings, long, tightly carried, not crossed.
Head: Slim, well rounded without knob and without dewlap.
Neck: Medium long, strong, upright and straight.
Legs: Medium long, strong, bright orange. Thighs, strong, well feathered, slightly showing. Feathers, smooth and flat with good down development.

Goose: Dual lobed. (Paunch in old birds acceptable.)

Colour

The blue

Plumage of both sexes: Light blue or grey colour to head, neck, chest, back, wings and thighs. The feathers of the shoulder, the wing and thighs show sharp white but not too wide lacing. Tail feathers are grey with white lacing. Belly and back silver-blue.

Eyes, large, dark brown framed with a narrow yellow ring. Beak, with black bean at tip and black serration (tooth-like edge) the remainder bright orange corresponding to colour of legs.

On the Continent there are also buff, cream, white and grey.

Weights

Gander 6–7 kg (13–15 lb)
Goose 5-6 kg (11–13 lb)

Scale of points

Size	10
Colour	20
Condition	10
Head and throat	20
Neck	5
Breast	5
Tail and paunch	5
Legs and feet	5
Carriage	20
	100

Defects

Body too heavy, curved neck, sign of knob, leg colour other than bright orange, missing black bean on beak, different beak or tooth colour, missing lacing.

TOULOUSE

Origin: France

France originated the Toulouse and developed it for table purposes. Stock was sent over to England and our breeders crossed the birds with our own English Greys, and developed the breed for body weight and quantity of flesh, as well as standard characteristics of plumage colour, markings and type.

General characteristics: male and female

Carriage: Thick set and somewhat horizontal, but not as upright in front as the Embden.
Type: Body long, broad and deep. Back slightly curved from the neck to the tail. Breast prominent, deep and full, the keel straight from stem to paunch, increasing in width to the stern and forming a straight underline. Shoulders broad. Wings large and strong. Tail somewhat short, carried high and well spread. Paunch and stern heavy and wide, with a full rising sweep to the tail.
Head: Strong and massive. Bill strong, fairly short and well set in a uniform sweep, or nearly so, from the point of the bill to the back of the skull. Eyes full.
Neck: Long and thick, the throat well gulleted.
Legs and feet: Legs short. Shanks stout and strong boned. Straight toes connected by web.
Plumage: Full and somewhat soft.

Colour

The grey

Plumage of both sexes: Neck grey. Breast and keel rather light grey, shading darker to thighs. Back, wings and thighs grey, each feather laced with an almost white edging, the flights without white. Stern, paunch and tail white, the tail with broad band of grey across the centre.

In both sexes: Bill, legs and webs orange. Eyes dark brown or hazel.

The buff

Identical to the grey in shape but buff to replace the grey colour requirements.

The white

To be pure white in plumage otherwise identical to grey.

Toulouse male

Weights

Gander	11.79–13.60 kg (26–30 lb)
Goose	9.07–10.88 kg (20–24 lb)

Scale of points

Size	20
Colour	10
Condition	10
Head and throat	15
Neck	5
Breast	10
Tail and paunch	10
Legs and feet	5
Carriage	15
	100

Defects

Patches of black or white in the plumage. Slipped or cut wings. Twisted keel.

OTHER BREEDS

The British Waterfowl Association looks after any other breeds of geese not standardized in this edition. New colour varieties of standardized breeds occur occasionally, but not in sufficient numbers to warrant standardizing.

West of England
A sex-linked breed with white males and grey and white females, medium in weight. Thought to have evolved through the traditional farmyard goose.

Shetland Goose
This is a smaller variety of the West of England Goose.

Ducks

It is generally accepted that all breeds of ducks, with the exception of the Muscovy, originated from the wild Mallard. This is quite clear with a breed like the Rouen, and some consider that the Black East Indian and the Cayuga originated from sports of the Mallard. It is possible to understand, too, the original white ducks of this country coming as Mallard sports. They may have been developed for body size and table qualities, by domestication and selection, resulting eventually in the Aylesbury as we know it today. In the make-up of the khaki Campbell the wild Mallard also played its part.

The British Waterfowl Association classifies the following as ornamental ducks; the Carolina, Mandarin and all other British or foreign species of wild duck.

ABACOT RANGER

Origin: Great Britain

The Abacot Ranger was developed primarily between 1917 and 1923 as a laying breed by Oscar Gray. Introduced into Germany in the early 1920s, a standard was drawn up in that decade that has not altered to the present day. A decorative and useful breed providing large white eggs and a good, well flavoured carcass.

General characteristics: male and female

Carriage: Slightly erect, alert and busy, drake a little more upright than duck.
Type: Body longish, well rounded. Long, almost straight back running approximately parallel with underside.

Colour

Drake's plumage: Head and neck brown–black overlaid with an intense green iridescence terminated shortly before the shoulders by a completely encircling silver-white ring. The ring dividing in sharp clean line the neck and breast colours. The breast neck-base, nape and shoulders a rich red-brown with silver-white lacing; the colour finishing in a line from the wing fronts. Belly, flanks and stern silvery white to cream, the white ground colour being preferable; a little of the breast colour washing along the upper flank permissible. Lower back dark grey with darker points and each feather laced with white. Rump brown-black with a slight iridescence, each feather lightly laced with white being desirable. The tail brownish-black, the whole bordered with white; brown-black 'sex curls'. The lower tail covert brown-black, finishing neatly and not running into the stern and flank colour. On the wings scapulars and tertiaries as the breast with wide silver-white lacing; wing bows (small coverts) French grey with a lighter edging; the violet/green speculum (wing bar) bordered by black and white lines on the upper and lower edges; primaries silvery white with a slightly iridescent dark grey overlay. Bill yellowish-green with black bean at the tip. Eyes dark brown. Legs and webs dark orange.
Duck's plumage: Head and neck fawnish-buff with the brow and crown strongly grained with dark brown (not black). Upper breast, lower neck, nape and shoulders lightly streaked with light brown on a pale cream ground. Lower breast, belly and stern creamy-white. Lower back light fawn with darker points and laced with light grey. Rump fawnish-grey with brown points forming a triangular pattern. On the wings scapulars and tertiaries creamy fawn with brown flecks; wing bows dark fawn laced with cream; the speculum and primaries as in the drake (the white ground colour to prevail). Tail is fawn. Bill dark grey, almost black. Eyes dark brown. Legs and webs as dark grey as possible.

Abacot Ranger male

Abacot Ranger female

In both sexes: The underdown is white and the duckling's down colour is golden-yellow.

Weights

Drake 2.50–2.70 kg ($5\frac{1}{2}$–6 lb)
Duck 2.25–2.50 kg (5–$5\frac{1}{2}$ lb)

Scale of points

Type	25
Size	20
Carriage	5
Colour (ground 10, bill 5, feet 5, head 10, wings 10)	40
Condition	10
	100

Defects

Drake: too dark a ground colour; breast colour running too low into the body; lack of lacing; brown head; broken, too wide or absent neck-ring.
Duck: coarse or black graining and an eye stripe on the head; absence of graining and streaking; white head; grey speculum (wing bars).
In both sexes: absent or defective wing bars and framing; yellow or blue bills. Keel in either sex.

AYLESBURY

Origin: Great Britain

The Aylesbury derives its name from the town of Aylesbury in Buckinghamshire. At the first poultry show of 1845, a class was provided for 'Aylesbury or other white variety', and another for 'Any other variety'. No doubt there were white ducks in this country for centuries before, and from them was developed by judicious selection for table purposes the white Aylesbury, which is today Britain's table breed *de luxe*. Once standardized as a breed it was developed by selecting for its distinctive characteristics, which separated it from all other white breeds of ducks.

General characteristics: male and female

Carriage: Horizontal, the keel parallel with the ground.
Type: Body long, broad and very deep, showing a good keel. Back straight, almost flat. Breast full and prominent. Keel straight from breast to stern. Wings strong and carried closely to the sides, fairly high but not touching across the saddle. Tail short, only slightly elevated, and composed of stiff feathers, the drake's having two or three well-curled feathers in the centre.
Head: Strong and powerful, with eyes as near the top of the skull as possible. Bill strong and wedge-shaped. When viewed from the side the outline is almost straight from the top of the skull, the head and bill measuring from 15–20 cm (6–8 in). Eyes full.
Neck: Curved and strong.
Legs and feet: Legs very strong and short, the bones thick, set to balance a level carriage. Feet straight and webbed.
Plumage: Bright and glossy, resembling satin.

Colour

Plumage of both sexes: White.
In both sexes: Bill pink-white or flesh. Eyes blue. Legs and webs bright orange.

Aylesbury male

Weights

Drake 4.55–5.44 kg (10–12 lb)
Duck 4.10–4.98 kg (9–11 lb)

Scale of points

Type	10
Size	20
Head and bill	20
Eyes	8
Keel	10
Colour	10
Neck	5
Legs and feet	5
Condition	12
	100

Defects

Plumage other than white. Bill other than white or flesh pink. Heavy behind. Any deformity.

BALI

Origin: East Indies

This is a breed currently staging a revival. It is closely related to the original ducks from Bali in shape and type.

Bali female

General characteristics: male and female

Carriage: Erect and active.
Type: Body slim and cylindrical with not too much shoulder. Wings packed tight to body. Legs strong, set well to rear of body.
Head: Rather Runner-like except for a small globular crest on the rear of the head.

Colour

Any colour permitted.

Weights

Drake	2.26 kg (5 lb)
Duck	1.81 kg (4 lb)

Scale of points

Body	30
Head and neck	20
Crest	10
Carriage and action	20
Colour	10
Condition	10
	100

Black East Indian male

BLACK EAST INDIAN

Origin: America

The Black East Indian is described in the first Book of Standards in 1865 but it has other names such as Buenos Aires, Labrador and Black Brazilian. Many consider this is a black sport from the Mallard.

General characteristics: male and female

Carriage: Lively, smart, symmetrical and clear of the ground from breast to stern. Slightly elevated.
Type: Body compact. Breast round and prominent.
Head: Neat and round, with high skull. Bill medium and fairly broad, well set in a straight line from the top of the eye. Eyes full.
Neck: Medium
Legs and feet: Legs of medium length. Feet straight and webbed.
Plumage: Bright and glossy.

Colour

Plumage of both sexes: A very lustrous, intense, beetle-green black, free from brown or white feathers.

In both sexes: Bill black. Eyes, legs and webs as black as possible.

Weights

Drake 0.90 kg (2 lb)
Duck 0.70–0.80 kg ($1\frac{1}{2}$–$1\frac{3}{4}$ lb)

Scale of points

Type	20
Size	20
Head, bill and neck	15
Legs and feet	5
Colour	30
Condition	10
	100

Defects

Purple sheen not desirable. Yellow or green bill a fault in either sex.

BLUE SWEDISH

Origin: Europe

The Blue Swedish duck has long been admired for its striking appearance. Its rich, well-laced colour, large size, fine length and carriage, with the added bonus of two white flight feathers, make this bird a challenge for any breeder.

General characteristics: male and female

Carriage: From 20° to 25° above the horizontal. Lively and alert.
Type: Body proportioned to dimensions of the back, round, plump, deep, without keel. Breast full, meaty, deep. Abdomen free of bulkiness, but round, full and capacious. Back flat, straight, about 50% longer than broad. Tail rather short, compact, carried slightly elevated. Wings neatly folded, maintained well up but not meeting over the back.
Head: Proportionate in size, oval, bold. Bill medium length. Eyes bold, bright.
Neck: Moderately tapering from shoulders. Faintly curved.
Legs and feet: Thighs well fleshed and of sufficient length to display the hocks just below the coverts. Shanks moderately short, strong.

Colour

Plumage of both sexes: Head in the drake dark blue with greenish reflections. In the duck same as body colour. Wings same as body colour except that the two outer primaries in each member should be white; speculums as inconspicuous as possible. Remainder of plumage a uniform shade of slate-blue, strongly laced with a darker shade of this blue throughout except for an unbroken, inverted heart-shaped 'bib', about 7.5 × 10 cm (3 × 4 in) in extent in the drake and 5.0 × 7.5 cm (2 × 3 in) in the duck upon the lower neck and upper breast. Bill in the drake blue preferred. In the duck blue-slate. Eyes brown. Legs and webs orange-black in drake, blue-brown in the duck.
Weights: Drake 3.6 kg (8 lb); duck 3.20 kg (7 lb)
Scale of points: Size 20; colour 25; condition 10; head, bill and neck 15; body and tail 15; legs and feet 5; carriage 10.

Defects

Lack of size, visible keel in either sex, white bib extending to lower mandible in duck. Russet tinge in plumage, black flecks in plumage.

CALL

Origin: Probably Holland. First standardized in Great Britain

These birds have been standardized since the first edition of this book when the breed was

Blue Swedish male

Blue Swedish female

known as the Decoy. Emphasis is placed up alertness in looks and movements.

General characteristics: male and female

Carriage: Carried well, and nearly level from breast to stern.
Type: Body very small and compact, broad and deep.
Head: Neat and round, with high skull. Bill short, maximum length 3.1 cm ($1\frac{1}{4}$in), and broad, set deep into the skull to give a square-looking appearance. Eyes full, round and alert.
Neck: Short.
Legs and feet: Legs short, set midway in the body. Feet straight and webbed.

Colour

The white

Plumage of both sexes: Pure white all over, free from sappy yellow colour.

In both sexes: Bill bright orange-yellow. Eyes blue.

Defects: Black specks on the duck's bill are a common minor fault with age. Any black in the drake's bill is a disqualification.

The grey or mallard

Drake's plumage: Head iridescent green with white collar almost encircling the neck. Breast rich purple-red with no white fringing. Back green/grey shading to black over the rump. Flanks light grey with black graining. Rump undercushion greenish-black. Bill green.
Duck's plumage: Head and neck golden-brown, each feather with darker brown graining. Faint eye stripes as in the wild mallard. Body almond-brown with dark brown concentric markings. Bill dark orange with brown saddle.

In both sexes: Speculum iridescent blue, edged with black then white.

Defects: White primaries, white feathers under the bill on the throat, or under the tail. Dark ground colour and lack of lacing in back of feathers of duck.

The pied

Coloured feathers as mallard. Variations on pied markings allowed but even markings to be encouraged.
Drake's plumage: Green head with white surround at base of bill and white line starting at rear of eyes and encircling back of head. Wide white ring encircling neck. Body as Mallard except underside of body behind legs white. Main tail feathers white. Outer flights white. Bill light yellow with greenish tinge.
Duck's plumage: Head marking as drake. Body white area often extends further. White in wings can extend onto shoulders. Bill yellow with light brown saddle.

The blue-fawn

Drake's plumage: Head charcoal-blue, breast claret, body light blue-grey shading to dark blue at the tail. Rest of body light grey with blue tinge. Bill colour green.
Duck's plumage: Head, the crown light blue with dark brown graining. Cheeks and front of neck light fawn. Faint eyelines, pale throat. Body feathers mainly blue. Bill colour light brown.

In both sexes: Flights smoke-grey, speculum matt charcoal-blue.

Defects: White primaries in either sex. Broad fawn lacing detaching from blue back feathers of duck.

The silver

Drake's plumage: Head as mallard but wider neck ring encircling the neck. Breast claret with each feather laced with white. Back light grey with black frosting, shading to black with beetle-green sheen on the rump. Some light claret feathers extend along the flanks. Belly white. Bill colour green.
Duck's plumage: Head and upper neck very pale fawn to cream or white with darker

Grey Call male

White Call female

Blue-fawn Call male

Blue-fawn Call female

graining on the crown. Body white with brown and grey mottling, more pronounced on upper body and breast. Bill colour light orange-brown with rich brown saddle.

In both sexes: Speculum iridescent blue.

Defects: Pied markings in head of drake, ducks with buff ground colour pencilled with brown.

The apricot

Drake's plumage: As Blue-fawn, but paler. Silvery grey head, breast light mulberry. Body light grey. Bill light green.

Duck's plumage: Rich apricot, some light grey on the back feathers. Flights light pearl-grey with apricot tinge. Speculum darker pearl-grey. Eye stripes on head. Bill light brown.

Defects: White primaries in either sex. In duck, neck ring, yellow bill in drake.

In both sexes and in all above colours

Webs to be orange, but can be slightly darker in the coloured Calls than the white Calls.

The magpie

Plumage in both sexes: Colour and markings as for the large magpie ducks.

The bibbed

Plumage in both sexes: Bodies and heads black, blue or lavender with white bibs. Bib as even as possible, an invented heart shape extending from the lower neck to the upper breast. White in outer wing flights (maximum of four flights). Bill colour in drake olive, in the duck black.

Bibbed and magpie: Legs dusky orange shaded irregularly with greyish-black.

Weights

Drake	570–680 g (20–24 oz)
Duck	450–570 g (16–20 oz)

Scale of points

Type	30
Size	20
Colour	20
Head, neck and bill	15
Legs and feet	5
Condition	10
	100

Note: Type takes precedence over colour.

Defects

Thin-bodied, boat-shaped bodies. Long, slim necks. Flat crowns, oval heads and square heads showing a flat top. Narrow cheeks, uneven cheeks. Long, narrow bills. Oversize. Long shanks.

CAMPBELL

Origin: Great Britain

It was the wild Mallard that played its part in the make-up of the khaki Campbell, together with blood of the fawn-and-white Runner and that of the Rouen. Introduced in 1901 by its originator Mrs Campbell of Uley, Gloucestershire, it was her special desire to keep the breed for prolific egg-laying, so that a very elementary standard was at first publicized. In this way the high egg-producing properties of the breed were maintained. The white Campbell came as a sport from the khaki. The dark Campbell was created by Mr H. R. S. Humphreys in Devon, to make sex-linkage in ducks possible.

General characteristics: male and female

Carriage: Alert, slightly upright and symmetrical, the head carried high, with shoulders higher than the saddle, and the back showing a gentle slant from shoulder to saddle; the whole carriage not too erect but not as low as to cause waddling. Activity and foraging power to be retained without loss of depth and width of body generally.
Type: Body deep, wide and compact, appearing slightly compressed, retaining depth throughout, especially from shoulders to chest and from middle of back through to thighs; broad and well-rounded front. Back wide, flat and of medium length, gently sloping with shoulders higher than saddle. Abdomen well developed at rear of legs, but not sagging; well-rounded underline of breast and stern. Wings closely carried and rather high. Tail short and small, rising slightly, the drake's with the usual curled feathers.
Head: Refined in jaw and skull. Face full and smooth. Bill proportionate, of medium length, depth and width, well set in a straight line with the top of the skull. Eyes full, bold and bright, showing alertness and expression, high in skull and prominent.
Neck: Of medium length, slender and refined, almost erect.
Legs and feet: Legs of medium length, and well apart to allow of good abdominal development, also not too far back. Feet straight and webbed.
Plumage: Tight and silky, giving a sleek appearance.
Quality and refinement: While aiming at good body size emphasis should be placed upon quality or refinement in general, i.e. neat bone, sleek silky plumage, smooth face, fine head points, etc. with absence of coarseness and sluggishness.

Colour

The dark

Drake's plumage: Head and neck beetle-green. Shoulders, breast, underparts and flank light brown, each feather finely pencilled with dark grey-brown, gradually shading to grey at stern close up to the vent, followed by beetle-green feathers with purplish tinge up to the tail coverts. Tail feathers dark grey-brown; coverts beetle-green with purplish tinge or reflection, also curled feathers in centre. Wing bow dark grey-brown laced with light brown; bar broad purplish-green, edged with a thin light grey line on each side; flights and secondaries dark brown.

Bill bluish-green with black bean-shaped mark at the tip. Eyes brown. Legs and feet bright orange.

Ducks plumage: Head and neck dark brown. Shoulders, breast and flank light brown, each feather broadly pencilled with dark brown, becoming brown towards stern, with lighter outer lacing, followed by beetle-green feathers at the rump. Back and wing bow dark brown, outer laced with lighter brown. Wing bar as in drake, but less lustrous. Tail and wing feathers dark brown.

Bill slaty brown with a black bean-shaped mark at the tip. Eyes brown. Legs and feet as near body colour as possible.

The khaki

Drake's plumage: Head, neck, stern and wing bar green-bronze. Remainder of plumage an even shade of warm khaki shading off to lighter khaki towards the lower parts of the breast.

Bill slate-blue preferred. Legs and webs dark orange.

Duck's plumage: An even shade of warm khaki. Head and neck a slightly darker shade of khaki. Breast lightly laced on close examination. Back laced. Wings khaki, sound top and under, wing bars brown.

Bill dark slate. Legs and webs as near the body colour as possible.

The white

Plumage of both sexes: Pure white throughout.

In both sexes: Bill, legs and webs orange. Eyes grey-blue.

Khaki Campbell male

Khaki Campbell female

White Campbell male

Weights

Drake 2.25–2.50 kg (5–5½ lb)
Duck in laying condition 2.00–2.25 kg (4½–5 lb)

Scale of points

Type (shape and carriage)	25
Size and symmetry	10
Head points	10
Legs and feet	5
Colour	25
Quality and refinement	15
Condition	10
	100

Defects

The dark: Yellow bill. White bib or white neck ring. Any deformity. Green eggs. Coarseness.
The khaki: Yellow bill. White bib or white neck ring. Streak from eyes in duck. White or light underpart or top of wings in either sex. White in wing bar. Any deformity. Excessive weight or coarseness. Lack of lacing in duck.
The white: Excessive weight or coarseness. Flesh-coloured bill. Any deformity.

Cayuga female

CAYUGA

Origin: America

Many consider that the Cayuga was bred from the Black East Indian along larger lines. The breed takes its name from Lake Cayuga, New York. It was in 1851 that black ducks made their appearance on the lake and specimens were sent to this country.

General characteristics: male and female

Carriage: Lively, clear of the ground from breast to stern.
Type: Body long, broad and deep. Breast prominent and forming a straight underline from stem to stern. Tail carried well out and closely folded, the drake's having two or three well-curled feathers in the centre.
Head: Large. Bill long, wide and flat, well set in a straight line from the tip of the eye. Eyes full.
Neck: Long and strong with a graceful curve.
Legs and feet: Legs large, strong boned, placed midway in the body, giving the bird a slightly elevated carriage. Feet straight and webbed.

Colour

Plumage of both sexes: A very lustrous green-black, free from purple or white.
In both sexes: Bill black. Eyes black. Legs and webs as black as possible.

Weight

Drake 3.60 kg (8 lb)
Duck 3.20 kg (7 lb)

Scale of points

Type	30
Size	20
Neck	5
Tail	5
Head and bill	10
Condition	10
Legs and feet	5
Colour	15
	100

Defects

Red or white feathers also brown or purple tinge. Orange coloured bill. Dished bill. Green coloured bill. Any deformity. Keel in either sex.

CRESTED

Origin: Great Britain

The Crested is somewhat like the Orpington in size and type. In breeding Crested ducks, not all the ducklings have crests, the plain being distinguishable from the crested at birth. Their utility qualities are comparable with those of the Orpington.

General characteristics: male and female

Carriage: Slightly elevated.
Type: Body long, broad and fairly deep. Full round breast, long broad back. Strong wings, carried closely. Short tail, similar to that of the Aylesbury.
Head: Long and straight. Crest globular, large, set evenly on the skull. Bill long and broad. Eyes large and bright.
Neck: Medium length. Slightly curved but not arched.
Legs and feet: Short and strong. Toes straight and connected by web.

Colour

Any colour is permitted.

Weights

Drake	3.20 kg (7 lb)
Duck	2.70 kg (6 lb)

Scale of points

Crest	25
Type	25
Size	15
Head and bill	10
Condition	15
Neck	5
Legs and feet	5
	100

Defects

Slipped wings. Twisted or roach back. Any deformity. Split crest.

White Crested male

Silver Appleyard Crested female

CRESTED MINIATURE

Origin: Great Britain

It was created in the British Isles during the late 1980s/early 1990s. In all points it is a miniature of the large Crested.

General characteristics and colour

As for large Crested.

Weights

Drake 1.125 kg ($2\frac{1}{2}$ lb)
Duck 0.90 kg (2 lb)

Scale of points

As for large Crested.

HOOK BILL

Origin: Almost certainly bred in the Netherlands in the seventeenth and eighteenth century. In old poultry books the North Holland White Breasted Duck has been called the ancestor of this breed. The curved bill was bred in to assist wildfowlers in recognizing the breed when in flight with wild Mallard returning from feeding grounds at night.

General characteristics: male and female

Carriage: Somewhat erect.
Type: Body medium length, fairly broad, breast round when viewed from front. Back long and slender, top line somewhat curved. Breast full, round and carried somewhat forward. Belly well developed, parallel to the back. Abdomen well developed but must not touch the ground. Wings strong and carried close to the body. Tail rather broad, follows the line of the back.
Head: When viewed from the front skull flat and no cheeks, with very little rise to the forehead. Bill flat, long and broad, strongly curved, when viewed with the head it is shaped like half a circle.
Neck: Vertical, rather long and thin.
Legs: Of medium length.

Colour

The white

A pure white throughout. Bill, white or flesh. Legs and webs bright orange.

The dark mallard

Drake's plumage: Head emerald-green. Neck dark green, at the lower end clear cut from breast colour without a white ring. Breast steel-blue. Body, upper part dark steel-blue, sides dark steel-blue with dark grey pencilling. Wings, wing bow, scapulars and tertiaries dark slate-blue; the secondaries brownish dark slate-blue; primaries dark brown-grey; the wing bars (speculums) dull brown with very little white bordering. Back, upper part ash colour with some green reflections, lower part green but not too shiny. Shoulders slate-blue with dark grey pencilled lines. Tail dark grey/brown edged with light brown. Tail covert and sex curls black with some reflections.
Duck's plumage: The markings, except for the eye lines and wing bars the same as in the Mallard coloured duck. The brown of the Mallard has been replaced by dark brown.

Hook Bill male

The duck has no stripes on either side of the head. The wings bars are dull brown with very little white edging.

White bibbed dark mallard

The same colour as the dark mallard colour except for a white heart-shaped bib upon the lower neck and upper breast. The outer wing primaries have three to six white feathers. Eyes grey/brown.

Bill colour in mallard varieties slate-grey with greenish tinge in drake. Slate-grey in duck.

Legs: Dark orange in both sexes.

Weights

Drake 2.25–2.50 kg (5–5½ lb)
Duck 1.75–2.25 kg (3¾–5 lb)

Scale of points

Type	25
Size	15
Head and bill	30
Condition	15
Neck	10
Legs and feet	5
	100

Defects

Short or straight bill. Any sign of keel. Over size.

INDIAN RUNNER

Origin: Asia

The era of the high egg-laying breeds of ducks started with the introduction of the Indian Runner into this country from Malaya. A ship's captain brought home fawns, fawn-and-whites and whites, distributing them among his friends in Dumfriesshire and Cumberland. They proved prolific layers, and there was a class of fawn Runners at the Dumfries show in 1876 but the fawn-and-whites were not exhibited until 1896. The Indian Runner Duck Club's Standard of 1907 described only the fawn-and-white, that of 1913 recognized also the fawn, while the 1926 Standard described the black and the chocolate varieties.

General characteristics: male and female

Carriage: Upright and active. The angle of inclination of the body to the horizontal varies from 50° to 80° when the bird is on the move and not alarmed; but when standing at attention, or excited, or specially trained for the show-pen, it may assume an almost perpendicular pose.

Type: Body slim, elongated and rounded, but slightly flattened across the shoulders. At the lower extremity the front line sweeps gradually round to the tail, which is neat and compact and almost in a line with the body or horizontally, but in some excellent birds slightly elevated or tilted upwards – the position of the tail varying with the attitude of the duck; however, habitually upturned sterns and tails (as in the Pekin duck) are considered objectionable. Stern short compared with other breeds, the prominence of the abdomen and stern varying in ducks according to the season and the age of the bird, being fuller when in lay; but a large pendulous abdomen and long stern, or a 'cut-away' abdomen and stern in young ducks should be avoided. Wings small in proportion to the size of the bird, tightly packed to the body and well tucked up, the tips of the long flights of the opposite wings crossing each other over the rump, more particularly when the bird is standing at attention. At the upper extremity the body gradually and imperceptibly contracts to form a funnel-shaped process, which again without obvious junction merges into the neck proper, the lower or thickest portion of this funnel-shaped process or 'neck expansion' being reckoned as part of the body.

Note: Total length of drake 65–80 cm (26–32 in) and duck 60–70 cm (24–28 in). Length of neck proper, from top of skull to where it joins the thick part of the 'funnel' about one-third the total length of the bird, not less. Measurements should be taken with the bird fully extended in a straight line, the bill and head in a line with the neck and body, and the legs and feet in the same straight line, the measurements being from the tip of the bill to the tip of the middle toe.

Head: Lean and racy looking, and with the bill wedge-shaped. Skull flat on top, and the eye socket set so high that its upper margin seems almost to project above the line of the skull. Eyes full, bright, very alert and intelligent. Bill strong and deep at the base where it fits imperceptibly into the skull. There should be no indication of a joint or 'stop'. The upper mandible should be very strong and nicely ridged from side to side, and the line of the lower mandible straight also, with no depression or hollow in the upper line from its tip to its base. The outline should run with a clean sweep from the tip of the bill to the back of the skull. The length and depth of the bill varies, but should never be out of balance or harmony with the rest of the head and the lines of the bird as a whole.

Neck: Long and slender, and when the bird is on the move or standing at attention, almost in a line with the body, the head being high and slightly forward. The thinnest part in fawn drakes is approximately where the dark bronze of the head and upper neck joins the lower or fawn of the neck proper. The muscular part should be well marked, rounded and stand out from the windpipe and gullet, the extreme hardness of feather helping to accentuate this. The neck should be neatly fitted to the head.

Legs and feet: Legs set far back to allow upright carriage. Thighs strong and muscular,

Black Indian Runner male

Fawn-and-white Indian Runner male

Trout Indian Runner male

White Indian Runner female

longer than in most breeds. Shanks short and feet supple and webbed. There should be sufficient width between the legs to allow free egg production, but not as much as to cause the duck, on actual test, to roll or waddle when in motion.
Plumage: Tight and hard.

Colour

The black

Plumage of both sexes: Solid black with metallic lustre like the Black East Indian. There should be no grey under the chin or wings, no grey wing ribbon, and no 'chain armour' on the breast.

In both sexes: Bill black. Legs and webs black or very dark tan.

The chocolate

Plumage of both sexes: A rich, solid chocolate throughout free from lacing. The drake, on assuming male plumage, darker than the duck, but the ground work is the same.

In both sexes: Bill, legs and webs black.

The fawn

Drake's plumage: Head and upper part of the neck dark bronze with metallic sheen, which may show a faint green tinge meeting the colour of the lower part of the neck with a clean cut or the lower colour merging into it imperceptibly. Lower neck and 'neck expansion' rich brown-red continued on to the breast, over the top of the shoulders and upwards to where it joins the head and upper neck colour, merging gradually on the back and breast into the body colour. Lower chest, flanks and abdomen French grey, made up of very minute and dense peppering of dark brown, or almost black, dots on a nearly white ground, giving a general grey effect without any show of white, the grey extending beyond the vent until it meets the dark or almost black feathers of the cushion under the tail. Scapulars (the long pointed feathers on each side of the back covering the roots of the wings) red-brown, peppered. Back and rump deep brown, almost black. Tail (fan feathers and curl) dark brown, almost black. Wing bow fawn, not pencilled; bar fawn corresponding with the coverts in the lower part, the upper part darker brown corresponding with the secondaries, which are black-brown with slight metallic lustre; primaries brown, fairly dark. When the drake is in 'eclipse' or in duck plumage he more closely approaches the duck in colour. All the dominant colours fade, but his head and neck are darker than the duck's; the body becomes a dirty fawn or ash, with perhaps some rustiness on the breast.

Bill pure black to olive green, mottled with black, and black bean. Legs and webs black, or dark tan, mottled with black.
Duck's plumage: The general plumage colour is an almost uniform warm ginger-fawn, with no marked variation of shade but having a slightly mottled or speckled appearance. When closely examined the head, neck, lower part of chest and abdomen may appear a shade lighter than the rest of the body. Each feather of the head and neck has a fine line of dark red-brown, giving a ticked appearance. Lower part of neck and neck expansion a shade warmer, each feather pencilled with a warm red-brown. Scapulars rich ginger-fawn, a shade darker than the shoulder and back, with well-marked red-brown pencilling. Wing bow a shade lighter than the scapulars but darkening towards the bar, the feathers pencilled as before; secondaries warm red-brown; primaries a shade lighter. Back and rump darker, the pencilling being richer and more marked, but the ground colour becoming lighter and warmer towards the tail. Tail lighter than upper parts of body, each feather pencilled. Belly lighter than upper parts of body, about the same shade of fawn as the head and neck, becoming a trifle darker on the tail-cushion, all feathers pencilled. In young ducks that have just completed the first adult moult there is often a rosy tinge on the lower neck expansion, upper part of breast and shoulders, but this soon fades away. Some ducks have a cream or light-coloured narrow band in the wing bar owing to the upper part of each feather being of a lighter or almost cream shade edged or laced again with the normal dark shade.

Bill black. Eye iris golden-brown. Legs and webs black or dark tan.

MAGPIE

Origin: Wales

The Magpie is a very striking duck with its boldly coloured and white plumage. It is, of course, a medium-sized breed which yields both meat and eggs, and is a good bird for the shows.

General characteristics: male and female

General shape and carriage: Fairly broad across and deep; good length of body, giving a somewhat racy appearance, indicative of strength combined with great activity.
Type: Back of great length, level and fairly broad across. Breast full and nicely rounded. Wings powerful, carried close to body. Tail medium length, gently rising from back, and increasing apparent length of bird; the drake having the usual curled feathers. Abdomen well developed.
Head: Long and straight. Eyes large and prominent, giving keen and alert appearance. Bill long and broad.
Neck: Long, strong and nicely curved.
Legs and feet: Medium, set wide apart. Feet straight and webbed.

Colour

Plumage of both sexes: Head and neck white, surmounted by a coloured cap covering the whole of the crown of the head to the top of the eyes. Breast white. Back solid colour from shoulders to the tip of the tail. White primary and secondary feathers. When the wings are closed there is a heart-shaped mantle of coloured feathers. Thighs and stern white. The Blue and White and Dun and White are marked as the Black and White with the blue and the dun replacing the black.

In both sexes: Bill yellow. In older birds, drake yellow spotted with green and duck grey/green. Eyes dark grey or dark brown. Legs and webs orange. Black on legs and webs a slight defect.

Weights

Drake 2.50–3.20 kg ($5\frac{1}{2}$–7 lb)
Duck 2.00–2.70 kg ($4\frac{1}{2}$–6 lb)

Scale of points

Weight	8
Condition	10
Head (bill, eyes)	10
Neck	2
Back	4
Tail	2
Wings	2
Breast	6
Body	20
Legs and feet	4
Colour	32
	100

Defects

Excessive weight or coarseness.

Magpie male

Magpie female

MUSCOVY

Origin: America

The Muscovy has also been known as the Musk duck, and the Brazilian. It is a distinct species, and wild ancestors of it were found in South America, which must have credit for the original source. Under domestication, our breeders have greatly increased the size of the ducks and also perfected their colour and markings.

General characteristics: male and female

Carriage: Low and jaunty.
Type: Body broad, deep, very long and powerful. Breast full, well rounded and carried low; keel long and well fleshed, just clear of ground, slightly rounded from stem to stern. Wings very strong, long and carried high. Tail long and carried low to give the body a longer appearance to the eye and a slightly curved outline to the top of the body.
Head: Large, and adorned with a small crest of feathers (more pronounced in the drake than duck) which are raised erect in excitement or alarm. Caruncles on face and over base of the bill. Bill wide, strong, of medium length and slightly curved. Eyes large, with wild or fierce expression.
Neck: Of medium length, strong and almost erect.
Legs and feet: Legs strong, wide apart and fairly short. Thighs short, strong and well fleshed. Feet straight, webbed, with pronounced toenails.
Plumage: Close.
Handling: Hard, well fleshed, muscular.

Colour

The white-winged black
Plumage of both sexes: A dense black throughout, except white wing bows. The black to carry a metallic-green sheen, or lustre, with bronze on the breast and parts of the neck.

The white-winged blue
Plumage of both sexes: Blue except for white wing bows.

The black
Plumage of both sexes: A dense beetle-green black throughout, with bronze on the breast and parts of the neck.

The white
Plumage of both sexes: Pure white throughout.

The blue
Plumage of both sexes: Light or dark shade of blue.

The black and white
Plumage of both sexes: Black and white, with defined regularity of markings.

The blue and white
Plumage of both sexes: Blue and white, with defined regularity of markings.

In both sexes and all colours
Bill yellow and black, red, flesh or lighter shade at point. Face and caruncles red or black. Eyes from yellow and brown to blue. Legs and webs white to black.

Weights

Drake 4.55–6.35 kg (10–14 lb)
Duck 2.25–3.20 kg (5–7 lb)

It is a characteristic of the breed for the drake to be about twice the size of the duck.

Lavender and white Muscovy male

Black and white Muscovy female

Scale of points

Shape and carriage	40
Head (including crest and carunculations)	20
Size	20
Condition	10
Colour	10
	100

ORPINGTON

Origin: Great Britain

It was from the blending of the Indian Runner, Rouen and Aylesbury that Mr W. Cook in Kent made the buff Orpington, intending it to be a dual-purpose breed. Its introduction followed that of the khaki Campbell, and it has been said that the originator was trying to make a strain of khaki duck. At one time very popular for its high laying qualities combined with table properties and also its beauty of plumage and colouring.

General characteristics: male and female

Carriage: Slightly elevated at the shoulders, but avoiding any tendency to the upright carriage of the Pekin or Indian Runner.
Type: Body long and broad, deep, without any sign of keel. Back perfectly straight. Breast full and round. Wings strong and carried closely to the sides. Tail small, compact and rising slightly from the line of the back, the drake's having two or three curled feathers in the centre. When in lay the duck's abdomen should be nearly touching the ground.
Head: Fine and oval in shape. Bill of moderate length, the upper mandible straight from bean to base in line with the highest point of the skull. Eyes large and bold, set high in the head. Deep set and scowling eyes are most objectionable.
Neck: Of moderate length, slender and upright.
Legs and feet: Legs of moderate length, strong and set well apart. Feet straight and webbed.
Plumage: Tight and glossy.

Colour

Drake's plumage: A rich even shade of deep red-buff throughout, free from lacing, barring and pencilling except that head and neck are seal brown with bright gloss, but complete absence of beetle-green, the seal brown to terminate in a sharply defined line all the way round the neck. The rump red-brown, as free as possible from 'blue'.
Duck's plumage: Similar to that of the drake's body, and free from blue, brown or white feathers.

Bill orange-brown with dark bean in duck. Bill yellow with dark bean in drake. Eyes brown iris and blue pencil. Legs and webs orange-red.

Weights

Drake 2.25–3.40 kg (5–7½ lb) min. and max.
Duck 2.25–3.20 kg (5–7 lb) min. and max.

Scale of points

Type and size	40
Head and eyes	10
Colour	40
Condition	10
	100

Buff Orpington male

Buff Orpington female

Defects

Colour other than stated. Twisted wings, wry tail, humped back or any other physical deformity. In the drake, grey, silver, or blue head, white feathers in neck, brown secondaries, beetle-green on any part, very green bill. In the duck, very heavy lacing, strong light line over the eyes, white feathers on neck or breast, brown feathers, green bill.

PEKIN

Origin: Asia

Bred in China, the Pekin reached this country around 1874 and about the same time stock also went to America. English breeders called for a plumage of buff/canary, sound and uniform, or deep cream, the former preferred. Nowadays it is customary to prefer the latter. In America, however, the Pekin became the producer of high-class table ducklings, with the standard plumage of 'creamy white'.

General characteristics: male and female

Carriage: Almost upright, elevated in front and sloping downwards to rear.
Type: Body broad and of medium length and without any indication of keel except a little between the legs. Back broad. Breast broad and full followed in underline by the keel (which shows very slightly between the legs) to a broad, deep paunch and stern carried just clear of the ground. Wings short, carried closely to the sides. Tail well spread and carried high, the drake's having two or three curled feathers on top. A good description of the general shape of the Pekin is it resembles a small wide boat standing almost on its stern, and the bow leaning slightly forward.
Head: Large and broad and round, with high skull, rising rather abruptly from the base of the bill, and heavy cheeks. Bill short, broad and thick, slightly convex but not dished. Eyes partly shaded by heavy eyebrows and bulky cheeks.
Neck: Short and thick, with slightly gulleted throat.
Legs and feet: Legs strong and stout, set well back and causing erect carriage. Feet straight and webbed.
Plumage: Very abundant, thighs and fluff well furnished with long, soft downy feathers.

Colour

Plumage of both sexes: Deep cream or cream.
In both sexes: Bill bright orange and free from black marks or spots. Eyes dark lead-blue. Legs and webs bright orange.

Weights

Drake 4.10 kg (9 lb)
Duck 3.60 kg (8 lb)

Scale of points

Type	25
Size	20
Head (bill 10, eyes 5)	15
Neck	5
Colour	15
Tail	5
Legs and feet	5
Condition	10
	100

Pekin male

Pekin female

Defects

Black marks or spots on bill. Any deformity.

ROUEN

Origin: France

Confirmation of the Mallard as the progenitor of duck breeds is found in the Rouen, which so closely resembles it in plumage markings, while the Rouen drake moults into duck plumage in the summer like the wild Mallard drake. The Rouen undoubtedly came from Rouen in France, and was known also as the Rhône duck. When first brought to England from France the breed was developed for table properties and was used in table crossings. Later it was bred, as today, for beauty of plumage and markings.

General characteristics: male and female

Carriage: Carried horizontal, the keel parallel with ground.
Type: Body long, broad and square. Keel deep. Breast broad and deep. Wings large and well tucked to the sides. Tail very slightly elevated, the drake's having two or three curled feathers in the centre.
Head: Massive. Bill long, wide and flat, set on in a straight line from the tip of the eye. Eyes bold.
Neck: Long, strong and erect, slightly curved but not arched.
Legs and feet: Legs of medium length. Shanks stout and well set to balance the body in a straight line. Feet straight and webbed.

Colour

Drake's plumage: Head and neck rich iridescent green to within about 2.5 cm (1 in) of the shoulders where the ring appears. Ring perfectly white and cleanly cut, dividing the neck and breast colours, but not quite encircling the neck, leaving a small space at the back. Breast rich claret, coming well under; cleanly cut, not running into the body colour, quite free from white pencilling or chain armour; chain armour or flank pencilling rich blue French grey, well pencilled across with glossy black, perfectly free from white, rust or iron. Stern same as flank, very boldly pencilled close up to the vent, finishing in an indistinct curved line (perfectly free from white) followed by rich black feathers up to the tail coverts. Tail coverts black or slate-black with brown tinge, with two or three green-black curled feathers in the centre. Back and rump rich green-black from between the shoulders to the rump. Wings: large coverts pale clear grey; small coverts French grey very finely pencilled; flight coverts dark grey or slate-black; bars (two, composed of one line of white in the centre of the small coverts) grey tipped with black, also forming a line at the base of flight coverts, the latter feathers slate-black on the upper side of the quill and rich iridescent blue on the lower side, each of these feathers tipped with white at the end of the lower side, forming two distinct white bars (the pinion bar being edged with black) with a bold blue ribbon mark between the two, each colour being clear and distinct and making a striking contrast; flights slate-black with brown tinge free from white. Markings throughout the whole plumage should be cleanly cut and well defined in every detail, the colours distinct and not shading into each other.

Bill bright green-yellow, with black bean at the tip. Eyes dark hazel. Legs and webs bright brick-red.

Duck's plumage: Head rich (golden, almond or chestnut) brown with a wide brown-black line from the base of the bill to the neck, and very bold, black lines across the head, above and below the eyes, filled in with smaller lines. Neck the same colour as the head, with a wide brown line at the back from the shoulders, shading to black at the head. Wing

Rouen male

Rouen female

bars, two distinct white bars with a bold blue ribbon mark between, as in the drake; flights slate-black with brown tinge, no white. Remainder of plumage rich (golden-almond or chestnut) brown of level shade, every feather distinctly pencilled from throat and breast to flank and stern, the markings to be rich black or very dark brown, the black pencilling on the rump having a green lustre.

Bill bright orange, with black bean at tip, and black saddle extending almost to each side and about two-thirds down towards the tip. Eyes dark hazel. Legs and webs dull yellow.

Weights

Drake 4.55–5.44 kg (10–12 lb)
Duck 4.10–4.98 kg (9–11 lb)

Scale of points

Drake

Type	10
Size	10
Head	5
Legs and feet	5
Colour (breast 10, bill 5, neck 5, chain armour 10, back and rump 5, wings 5, stern 5, tail 5)	50
Markings	10
Condition	10
	100

Duck

Type	10
Size	10
Head	5
Legs and feet	5
Colour (ground 15, bill 10, head 5, neck 5, wings 5)	40
Pencilling	20
Condition	10
	100

Defects

Leaden bill. No wing bars. White flights. Stern broken down. Wings down or twisted. Any deformity. In the drake, black saddle, black bill or minus ring (on neck); in the duck, white or approaching white ring (on neck).

ROUEN CLAIR

Origin: France

It must remind one of the common Mallard – a body long and developed, the width corresponding with the length.

General characteristics: male and female

General shape and carriage: Very long, graceful with good width. The important feature of the Rouen Clair is its length (ideally 90 cm (35 in) from point of beak to end of tail with neck extended), which gives it elegance in spite of its heavy body.
Type: Rather more upright and smooth breasted than the Rouen.
Head: As for Rouen.

Rouen Clair male

Rouen Clair female

Neck: As for the Rouen.
Legs and feet: As for the Rouen.

Colour

Drake's plumage: Head and neck green with clear white collar, covering about four-fifths of the neck – no grey in the green. Breast (bib) red-chestnut, with light white borders at the end of each feather. Belly light grey, changing to white. The underneath of the tail is black. Back, the above pearly grey, darker than the flanks. Wings violet-blue (indigo) with brilliant reflecting powers, the ends of which above and below are limited by a white streak. Sides (flanks) pearly grey – no mixture of chestnut feathers. The stern is white. Rump, brilliant black. Tail feathers whitish-grey, garnished with some curled-up feathers.

Bill yellow, with a slight greenish tint and no black lines along its centre. Eye yellow. Legs and webs yellow-orange.

Duck's plumage: Wings to be the same as the drake with the brilliant reflections. Eyebrows, above the eye presenting a slight curve nearly white which forms the eyebrow. On the cheek another white line going from the eye to the start of the bill. Feathers, the ground colour should be of a dark fawn on the back marked with a single 'V' shape on each feather in dark brown. The flanks although of a lighter shade of fawn should be marked the same. Each breast feather should be delicately ticked with dark brown markings on a light brown ground. The top of the bill and the front of the neck are of a pale colour (creamy shade extending not too low on the breast).

Bill colour bright orange-ochre with brown saddle. Eye, pupil blue, iris very dark brown. Legs and webs yellow-orange.

Weights

Drake 3.40–4.10 kg ($7\frac{1}{2}$–9 lb)
Duck 3.00–3.40 kg ($6\frac{1}{2}$–$7\frac{1}{2}$ lb)

Scale of points

Size	8
Condition	10
Head, eyes and bill	8
Neck	6
Back	4
Breast	6
Belly	2
Tail	2
Wings	2
Legs	2
Feet	2
Breed character	10
Colour	38
	100

Defects

Keel in either sex, lack of eye stripes in duck, lack of white stern in drake.

SAXONY

Origin: Germany

The Saxony duck made its first public appearance in the Saxony County Show of 1934. It

Saxony male

Saxony female

is a dual-purpose domestic duck, attractive in appearance, producing full breasted, meaty birds and not less than 150 eggs per year.

General characteristics: male and female

Carriage and appearance: Strong, with a long broad body and no trace of keel.
Type: Body long and stocky. Back broad and long, sloping slightly to the rear. Breast broad and deep without a keel. Wings not too long, lying into the sides. Tail long, carried full and closed.
Head: Long and flat. Bill average length and broad.
Neck: Average length, not thin.
Legs and feet: Average length, set almost in the middle of the body.
Plumage: Lies close into the body.

Colour

Drake's plumage: Head and neck blue as far as the white neck ring which completely encircles the neck. Lower part of the neck, shoulders and breast rusty red with slight silver lacing on shoulders and breast. Back and rump blue-grey. Lower body, flights and main tail oatmeal.
Duck's plumage: Head, neck and breast buff with a white eye line above the eye and also a lower one in front of the eye. Cream throat. Back and breast buff. Rump buff. Flights oatmeal, wing bars grey.

Weights

Drake	3.60 kg (8 lb)
Duck	3.20 kg (7 lb)

Scale of points

Type	20
Size	20
Head, bill and neck	15
Eyes	5
Colour	25
Legs and feet	5
Condition	10
	100

Defects

Broken neck ring, brown head, black bean on bill, breast colour running onto the flanks of the drake. Pale neck ring, absence of eye stripes, white bib, dark underfeathering in the duck. Pale eyes, upright walk, slipped wing. Keel in either sex.

SILVER APPLEYARD

Origin: Great Britain

The Silver Appleyard is a good all-round utility duck produced by Mr Reginald Appleyard from selective cross-breeding. It is a good layer, an excellent table bird and very ornamental.

General characteristics: male and female

Carriage: Lively, slightly erect, the back sloping gently from shoulder to tail.

Silver Appleyard male

Silver Appleyard female

Type: Body compact, broad and well rounded. Tail broad, the drake having the usual curled tail feathers.
Head: Bold and alert.
Neck: Upright and of medium length.
Legs and feet: Average length, set slightly back.

Colour

Drake's plumage: Head and neck are black-green with silver-white flecked throat, faint eyebrow and cheek markings in silver. Silver-white ring completely circling the base of the neck. Base of neck and shoulders below the ring light claret. Silver light wing coverts matching the breast and underbody, followed by the usual band of iridescent blue. Back and rump black-green with white tips to tail feathers.
Duck's plumage: Head, neck and underbody silver-white with crown and back of neck flecked with fawn. Deep fawn line through eyes. Shoulders and back strongly flecked with fawn, with the usual iridescent blue in the wings. Tail fawn.

In both sexes: Bill yellow. Eyes dark hazel. Legs orange.

Weights

Drake 3.60–4.10 kg (8–9 lb)
Duck 3.20–3.60 kg (7–8 lb)

Scale of points

Type	30
Colour	25
Size	20
Carriage	15
Condition	10
	100

Defects

Lack of cheek and throat markings. Absence of blue wing bars. Keel in either sex.

SILVER APPLEYARD MINIATURE

Origin: Great Britain

These are a miniature of the large Silver Appleyard, with general characteristics and colour as for the large.

Weights

Drake 1.36 kg (3 lb)
Duck 1.19 kg ($2\frac{1}{2}$ lb)

Scale of points

Type	30
Colour	25
Size	20
Carriage	15
Condition	10
	100

Silver Appleyard Miniature male

Silver Appleyard Miniature female

SILVER BANTAM DUCK

Origin: Great Britain

This bantam was formerly known as Silver Appleyard bantam although the colouring was not the same as the illustration in the previous Standards Book showed.

General characteristics: male and female

Carriage: Sprightly, slightly erect, head held high, the back sloping gently from shoulder to saddle. Keel well clear of the ground, for active foraging.
Type: Body compact with width well maintained from stem to stern. Tail short and small, drakes have the usual curled tail feathers.
Head: Fine-drawn and held high.
Neck: Almost vertical, medium in length.

Colour

Drake's plumage: Head and neck black with green sheen. Breast and shoulders red-brown with white lacing. Belly, flank and stern silver-white. Back grey with black stippling finishing in black on rump. Wings, scapulars as shoulders, violet-green speculum; flights white with grey stippling on outer edges. Bill yellow-green.
Duck's plumage: Head and neck fawn with graining of dark brown or black. Breast and shoulders cream with brown streaks. Lower breast, belly and stern cream. Rump fawn-grey with brown flecks. Tail fawn. Wing bows dark fawn with cream lacing; speculum as drake; flights as drake.

Bill yellow to grey-green with black markings.

In both sexes: Eyes dark brown. Legs orange.

Weights

Drake	900 g (2 lb)
Duck	850 g ($1\frac{3}{4}$ lb)

Scale of points

Type	20
Size	15
Head, bill and neck	15
Carriage	10
Colour	20
Legs and feet	5
Condition	15
	100

WELSH HARLEQUIN

Origin: Great Britain

General characteristics: male and female

Alert, slightly upright and symmetrical, the head carried high, with shoulders higher than saddle and the back showing a gentle slant from shoulder to saddle; the whole carriage not too erect but not as low as to cause waddling. Activity and foraging power to be retained without loss of depth and width of body generally.
Type: Body compact with width well maintained from stem to stern. Tail short and small. Drakes have the usual curled tail feathers.

Welsh Harlequin male

Welsh Harlequin female

Head: Refined in jaw and skull. Bill proportionate, of medium length, depth and width, well set in a straight line with the top of the skull. Eyes full, bold and bright, showing alertness and expression, high in skull and prominent.
Neck: Of medium length, slender and refined, almost erect.
Legs and feet: Of medium length and well apart to allow good abdominal development; not too far back. Feet straight and webbed.
Plumage: Tight and silky, giving a sleek appearance.

Colour

Drake's plumage: Head and upper neck iridescent green overlaid with bronze lustre, to within about 2.5 cm (1 in) of the shoulders where a 1.5 cm ($\frac{1}{2}$ in) wide white ring completely encircles the neck. The ring is of sharp definition. The breast, neck base and shoulders should be rich red/brown/mahogany finely laced with white; the colour washes along the flanks, finishing at the thigh coverts. The belly and stern are creamy white. The lower back becomes slaty/grey as it meets the rump, which is the same beetle-green as the head. The tail is dark brown, finely bordered all round with white and carries the usual 'sex-curls' in the colour of the rump. The lower tail covert is also beetle-green in colour. The wings, scapulars and tertiaries are as the breast colour, laced with creamy white and giving a rich tortoiseshell effect; wing bows light mahogany laced with creamy white; the speculum is bronze/green bordered by lines of black and white on upper and lower edges; primaries white overlaid with brown.

The bill is olive-green without any trace of blue, with a black bean at the tip. Eyes dark brown. Legs dull orange.

Duck's plumage: Head and upper neck honey/fawn with brown graining on crown. Shoulders, breast, belly, flank and stern fading through fawn to creamy white. Back and wings a mixture of fawn, red/brown and cream producing a rich tortoiseshell effect, with well-defined lacing on the tertiaries; the wing bars are bronze and the primaries brown edged with white. The rump is mid-brown with darker brown centres to each feather; the tail is mid-brown.

Bill gun metal/slate grey. Eyes dark brown. Legs dark brown.

Weights

Adult drake	2.25–2.50 kg (5–5$\frac{1}{2}$ lb)
Adult duck	2.00–2.25 kg (4$\frac{1}{2}$–5 lb)

Scale of points

Type	20
Size	15
Head, bill and neck	15
Carriage	10
Colour	20
Legs and feet	5
Condition	15
	100

Defects

Blue wing bars. Lack of neck ring in drake. Bill blue on drake. Yellow bill on duck, lack of feather markings on rump. Oversize in either sex.

OTHER BREEDS

The British Waterfowl Association looks after any other breeds of duck not standardized in this edition. Specimens of other breeds appear occasionally, but not so far in sufficient numbers to warrant standardizing.

Pomern
A small Blue Swedish without white flights, Eastern European in origin.

Shetland
A small and less well marked version of the Blue Swedish, but black where the Swedish is blue.

Standard for eggs

The Poultry Club has authorized the following standard and scale of points for judging eggs.

EXTERNAL

Shape: Showing ample breadth, good dome, with greater length than width, the top to be much roomier than the bottom and more curved. The bottom should not be too pointed, and a circular, or even narrow shape is undesirable. The ideal shape is described as an elliptical cone. In outline it is an asymmetrical ellipse or 'Cassinian oval' and a cross-section at any point across the egg's girth is a perfect circle. This description is best shown by the large fowl egg. Pullet eggs are less pointed whereas some breeds of bantams characteristically lay more pointed eggs.

Turkeys, ducks and geese are distinct species and each lays eggs of slightly different shape. Hence, they should be shown in their own classes. Turkey eggs are quite short and conical. Duck eggs are slightly elongated and those of bantam ducks tend to be pointed. Geese lay eggs which are lacking for girth and narrow towards the pointed end.

Size: Mere size is not a deciding point but should be appropriate for the breed and species. A pullet's normal egg when the bird starts to lay is 49.6 g ($1\frac{3}{4}$ oz) and increases quickly to 56.7 g (2 oz), exceeding that after several months of production. There is another increase in the hen egg after the moult. Bantam eggs should not exceed 42.5 g ($1\frac{1}{2}$ oz). Eggs weighing in excess of this should be passed.

Turkey and duck eggs weigh between 70.9 g ($2\frac{1}{2}$ oz) and 92.2 g ($3\frac{1}{4}$ oz). Bantam duck eggs should not exceed 63.8 g ($2\frac{1}{4}$ oz). Goose eggs vary with breed. Light geese lay eggs from 141.8 g (5 oz) and heavy breed goose eggs can weigh up to 198.6 g (7 oz).

Shell texture: Smooth, free from lines or bulges, evenly limed, smooth at each end, without roughness, porous parts or lime pimples.

Colour: White, cream, light brown (tinted), brown, mottled or speckled, blue, green, olive and plum. The colour should be even and in the case of mottled or speckled eggs, regular mottles or speckles are preferred. Mottled or speckled eggs are shown according to their ground colour. Where a Breed Club has stated in its standard that a particular colour is required, any variations from this should be penalized.

Freshness, bloom and appearance: Shells to be clean, without dull or stale appearance as befits a new-laid egg. Shell surfaces may be shiny or matt, but should be free from blemishes such as stains and nest marks. Eggs may be washed in preparation but not polished.

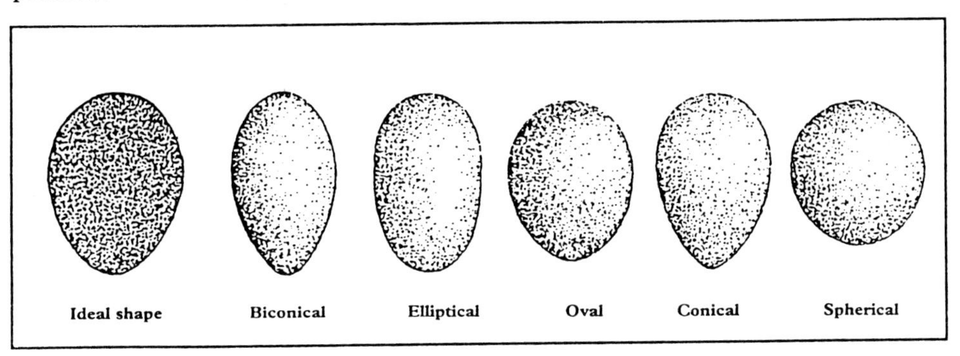

(*a*) **Large fowl: single white egg**

(*b*) **Large fowl: single dark brown egg**

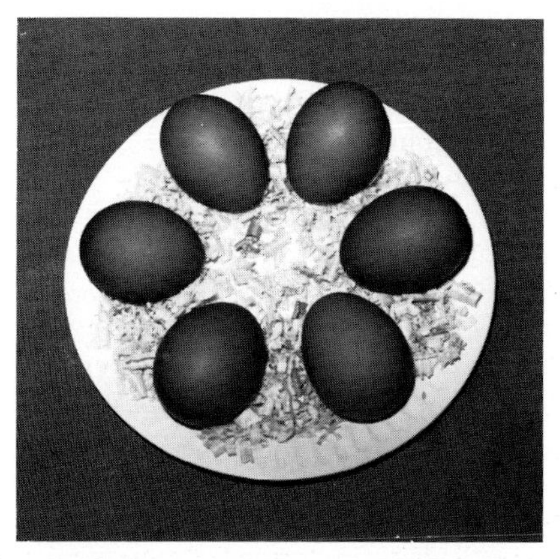

(*c*) **Bantam: six dark brown eggs**

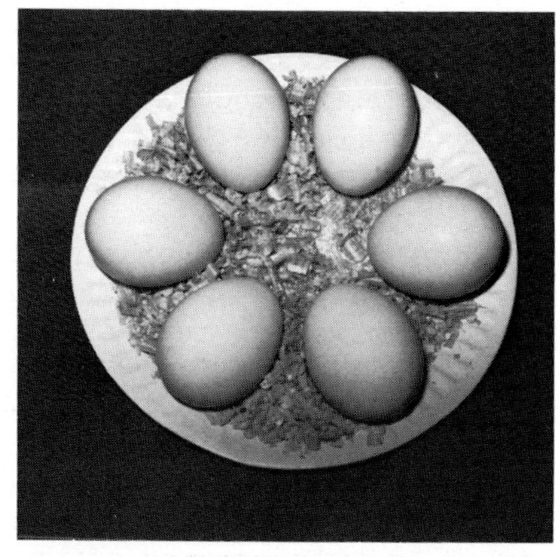

(*d*) **Large fowl: six white eggs**

(*e*) **Bantam: three distinct egg colours**

(*f*) **Bantam: three cream eggs**

British Waterfowl Association: three waterfowl eggs

Bantam: three light brown eggs

In duck eggs the position of the air space can be apparent. This is not considered a fault. Muscovy duck eggs often have a wax cuticle which may be removed.
Matching and uniformity: Eggs forming a plate or exhibit to be uniform in shape, shell texture, size, colour and appearance.

INTERNAL

Yolk: Rich, bright golden-yellow, free from blood streaks or 'meat' spots. Well-rounded and well-raised from the centre of the albumen. One uniform shade. Blastoderm or germ spot not discoloured and there should be no sign of embryo development.
Albumen: This is clear with no signs of blood spots or cloudiness and preferably with no tint of colour. It is of dense substance, particularly around the yolk and the differentiation between this thick albumen and the thin outer should be distinct. Waterfowl albumen must be clear, it is also more viscous and distinct than the hen's albumen and for these reasons waterfowl contents should be exhibited in classes separate from large fowl and bantam.
Chalazae: Each chalaza to resemble a thick cord of white albumen opposite each other and attached to the yolk, keeping it to the centre of the inner albumen. Free of blood and 'meat' spots.
Airspace: Small, about 1.5 cm ($\frac{1}{2}$ in) diameter (1 cm ($\frac{3}{16}$ in) bantams), the membrane adhering to the shell. It should be placed at the broad (dome) end, ideally just to one side.
Freshness: Indicated by small, taut airspace and unwrinkled top surface of yolk which should be raised and not lacking in height. A stale albumen lacks differentiation and is watery and runny.

Scale of points

External	
Shape	25
Size	15
Shell texture	20
Colour	20
Freshness, bloom and appearance	20
	100*

* May be maximum for each egg, or for a plate of eggs, whatever the number. Add 5 points more for each egg for matching and uniformity.

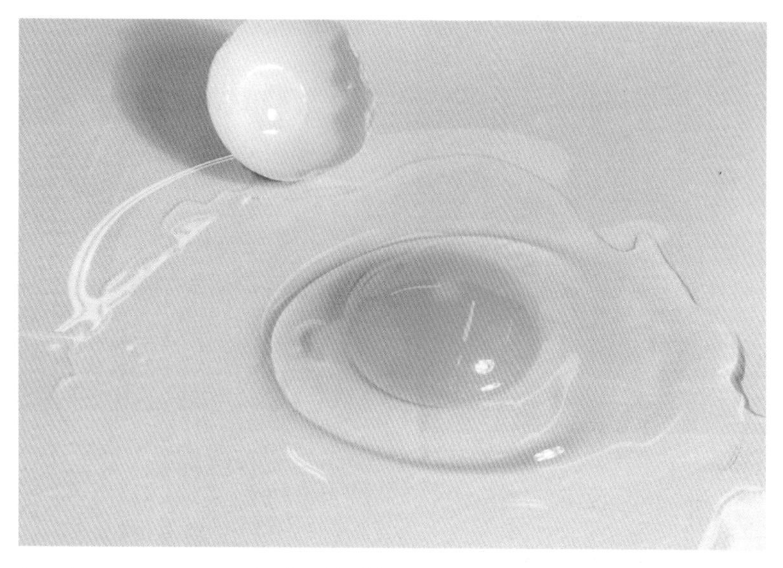

Fresh egg contents: raised yolk,thick albumen and small air sac

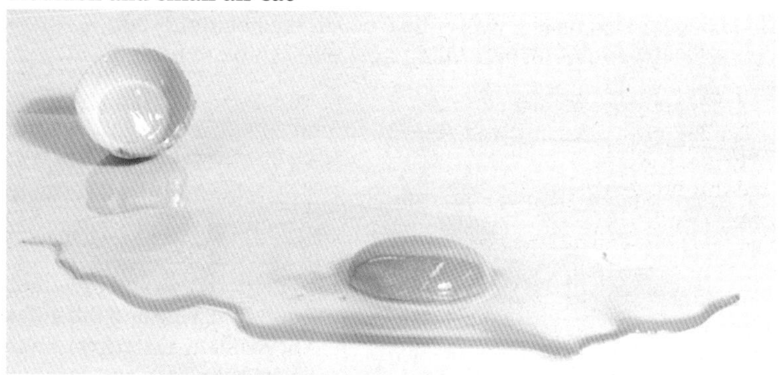

Stale egg contents: flat yolk, thin albumen, large air sac and blood spots

Internal	
Yolk	30
Albumen	30
Chalazae	10
Airspace	10
Freshness	20
	100

Serious defects (for which eggs *should* be passed)

More than one yolk. Staleness. Polished or over-prepared shells. Overweight in bantam eggs including contents classes. A developing embryo as shown by a 'halo' around the germ spot. Excessive blood streaks and 'meat' spots.

Disqualification

Addition of colouring to shells. Artificial polish or colouring would amount to disqualification and a report to the Poultry Club of Great Britain.

GLOSSARY

Abdomen Underpart of body from keel to vent.
A.O.C. Any other colour.
A.O.V. Any other variety.
Axial feather Small feather between wing primaries and secondaries (not waterfowl).

Back Top of body from base of neck to beginning of tail.
Bands Stripes straight across a feather (see also 'Cuckoo' and 'Pencilling').
Bantam Miniature fowl, formerly accepted as one-fifth the weight of the large breed it represented, but nowadays one-fourth.
Barring Alternate stripes of light and dark across a feather, most distinctly seen in the barred Plymouth Rock.
Bay A reddish-brown colour (see also 'Wing bay').
Beak The two horny mandibles projecting from the front of the face.
Bean A black spot or mark (generally raised) at the tip of the upper mandible of a duck's bill, seen in Cayugas and other breeds of waterfowl.
Beard A bunch of feathers under the throat of some fowls, such as the Faverolles, Houdan and some varieties of Poland. A tuft of coarse hair growing from the breast of an adult turkey male, also known as the 'tassel'.
Beetle brows Heavy overhanging eyebrows, best seen in the Malay.
Bill A duck's beak.
Blade Rear part of a single comb.
Blocky Heavy and square in build.
Booted Feathers projecting from the shanks and toes, as in the Brahma, Cochin, and Booted bantam.
Boule Hackle starting at sides of throat with a tendency to join behind the neck to form a mane, very thick and convexly arched, reaching to shoulder and saddle and covering the whole back. Seen in some Belgian bantams and Orloffs.
Bow legged Greater distance between legs at the hocks than at knees and feet.
Brassiness Yellowish foul colouring on plumage, usually on back and wing.
Breast Front of a fowl's body from point of keel bone to base of the neck. In dead birds, flesh on the keel bone.
Breed A group of birds answering truly to the type, carriage and characteristics distinctive of the breed name they take. There may be varieties within a breed, distinguished by differences of colour and markings.

Cap A comb; also the backpart of a fowl's skull; head markings as seen in some ducks.
Cape Feathers under and at base of neck hackle, between the shoulders.
Capon Strictly speaking a castrated male fowl, but term is also used to describe one treated chemically.
Carriage The bearing, attitude, or style of a bird, especially when walking.
Caruncles Fleshy protuberances on head and wattles of turkeys and Muscovy ducks.
Chicken A term employed by the Poultry Club to describe a bird of the current season's breeding.
Cinnamon A dark reddish-buff colour.
Cloddy see 'Blocky'.
Cloudy Indistinct (see 'Mossy').
Cobby Short, round or compact in build.
Cock A male bird after the first moult.
Cockerel A male bird of the current year's breeding.
Cockerel-breeder A term applied to birds, either male or female, selected to produce good standard bred cockerels.
Collar A white mark almost encircling the neck of the Rouen drake, also known as the 'ring'.
Comb Fleshy protuberance on top of a fowl's head, varying considerably in type and size and including cushion (Silkie), horn or V-shaped (La Flèche and Sultan), leaf or shell (Houdan), pea or triple (Brahma), rose (Hamburgh, Wyandotte, etc.), single (Cochin, Leghorn, etc.), cup (Sicilian Buttercup), strawberry or walnut (Malay) and raspberry (Orloff).

Concave sweep Hollow curve from shoulders to part way up the tail.
Condition State of a bird's health, brightness of comb and face and freshness of plumage.
Corky Light but firm handling characteristic in O.E.Game.
Coverts Covering feathers on tail and wings.
Cow hocks Weakness at hocks (see 'Knock-kneed').
Crescent Shaped like the first or last quarter of the moon.
Crest A crown or tuft of feathers on the head; known also as 'top knot' and in Old English Game as the 'tassel'.
Crow head Head and beak narrow and shallow, like a crow.
Cuckoo banding Irregular banding where the two colours are somewhat indistinct and run into each other, as in the cuckoo Leghorn and Marans.
Cup comb A comb somewhat resembling a tea-cup with the edges spiked as in the Sicilian Buttercup.
Cushion A mass of feathers over the back of a female covering the root of her tail, and most prominently developed in the Cochin.
Cushion comb An almost circular cushion of flesh, with a number of small prominences over it, and having a slight furrow transversely across the middle, as in the Silkie.

Daw eyed Having pearl-coloured eyes (jackdaw).
Deaf-ears see 'Ear-lobes'.
Dewlap The gullet (so-called), seen to the best advantage in adult Toulouse geese. Loose pouch of skin on throat under the beak.
Diamond The wing bay. A term commonly used among Game fanciers.
Dished bill Depression or hollow in the upper line of the bill of a duck or drake.
Dished lobe Lobe that is hollow in the centre.
Double comb see 'Rose comb'.
Double laced Two lacings of black as on an Indian Game female's feather. First there is the outer black lacing round the edge of the feather and next the inner or 'second' lacing (see 'Lacing').
Down Initial hairy covering of baby chick, ducklings, etc. Also the fluffy part of the feather below the web and small tufts sometimes seen as faults on toes and legs of clean legged breeds (see 'Fluff').
Drake Male duck.
Dual lobed see 'Lobe'.
Dubbing Removal of comb, wattles, and ear-lobes, so as to leave a fowl's head smooth.
Duck General term for certain species of waterfowl, and also used to describe the female.
Duck footed Fowls having the rear toe lying close to the floor instead of spread out, thus resembling the foot of a duck.
Dusky Yellow pigment shaded with black.

Ear-lobes Folds of skin hanging below the ears, sometimes called 'deaf-ears'. They vary in size, shape, and colour, the last named including purple-black, turquoise-blue, cream, red, and white.

Face The skin in front of, behind, and around the eyes.
Feather legged Characteristic of breeds such as the Brahma, Cochin, Faverolles, etc. May be sparsely feathered down to the outer toes, as in the Faverolles, or profusely feathered to the extremity of middle and outer toes as in the Brahma. Serious defect in clean legged breeds.
Flat shins Shanks that are flat fronted instead of rounded.
Flight coverts Small stiff feathers covering base of the primaries.
Flights see 'Primaries'.
Fluff Soft downy feathers around the thighs, chiefly developed in birds of the Cochin type; the downy part of the feather (the undercolour) not seen as a rule until the bird is handled; also the hair-like growth sometimes found on the shanks and feet of clean legged fowls, and in this case a defect.
Footings see 'Booted'.
Foxy Rusty or reddish in colour (see also 'Rust').
Frizzled Curled; each feather turning backwards so that it points towards the head of the bird.
Furnished Feathered and adorned as an adult. A cockerel that has grown his full tail, hackles, comb, etc. is said to be 'furnished'.

Gander The male of geese.
Gay Excess white in markings of plumage.
Goose The female of geese.
Grizzled Grey in the flights of an otherwise black bird.
Ground colour Main colour of body plumage on which markings are applied.

Gullet The loose part of the lower mandible; the dewlap of a goose. It appears on fowls, and is seen most distinctly perhaps on old Cochin hens, when it resembles a miniature beard of feathers.
Gypsy face The skin of the face a dark purple or mulberry colour.

Hackles The neck feathers of a fowl and the saddle plumage of a male consisting of long, narrow, pointed feathers.
Hangers Feathers hanging from the posterior part of a male fowl – the lesser sickles and tail coverts known as tail hangers, and the saddle hackle as saddle hangers.
Hard feather Close tight feathering as found on Game birds.
Head Comprises skull, comb, face, eyes, beak, ear-lobes and wattles.
Hen A female after the first adult moult.
Hen feathered A male bird without sickles or pointed hackles (sometimes called a 'henny').
Hind toe The fourth or back toe of a fowl.
Hock Joint of the thigh with the shank, sometimes called the knee or the elbow.
Hollow comb Depression in comb.
Hollow lobes Depression in ear-lobes.
Horn comb A comb said to resemble horns, but generally similar to the letter V, and seen to the best advantage on a matured La Flèche or Sultan male. The comb starts just above the beak, and from it branch two spikes thick at the base and tapered at the end.

In-kneed see 'Knock-kneed'.
Iris Coloured portion of eye surrounding the pupil.

Keel Blade of the breastbone; in ducks the dependent flesh and skin below it. In geese the loose pendant fold of skin suspended from the underpart of the bone.
Keel bone Breastbone or sternum.
Knob Protuberance on upper mandible of certain brands of geese.
Knock-kneed Hocks close together instead of well apart.

Lacing A stripe or edging all round a feather, differing in colour from that of the ground; single in such breeds as the Andalusian, Wyandotte, and Sebright bantams, and double in Indian Game and other females. In the last case the inner lacing not as broad as the outer (see also 'Double laced').
Leader The single spike terminating the rose type of comb; also know as the 'spike'.
Leaf comb A comb resembling the shape of a butterfly with its wings nearly wide open, and the body of the insect resting on the front of the fowl's head. It has also been referred to as resembling two scallop shells joined near the base, the join covered with a piece of coral. Seen to the best advantage on a Houdan male.
Leg The shank or scaly part.
Leg feathers Feathers projecting from the outer sides of the shanks of such breeds as the Brahma, Cochin, Faverolles, Langshan and Silkie.
Lesser sickles see 'Sickles'.
Lobe (see 'Ear-lobes' for fowl) In waterfowl the pouch in between the bird's legs. Dual lobed means two pouches, single lobed, one central pouch.
Lopped comb Falling over to one side of the head.
Lustre see 'Sheen'.

Main tail feathers see 'Tail feathers'.
Mandibles Horny upper and lower parts of beak or bill.
Marking The barring, lacing, pencilling, spangling, etc. of the plumage.
Mealy Stippled with a lighter shade, as though dusted with meal, a defect in buff-coloured fowls.
Moons Round spangles on tips of feathers.
Mossy Confused or indistinct marking; smudging or peppering. A defect in most breeds.
Mottled Marked with tips or spots of different colour.
Muff Tufts of feathers on each side of the face and attached to the beard, seen in such breeds as the Faverolles, Houdan, and some varieties of Poland; also known as 'whiskers'.
Muffling The beard and whiskers, i.e. the whole of the face feathering except the crest. In Old English Game the muffed variety has a thick muff or growth of feathers under the throat, differing in formation from that of the breeds named under 'Muff'.
Mulberry see 'Gypsy face'.

Nankin see breed for colour description.

Open barring Where the bars on a feather are wide apart.
Open lacing Narrow outer lacing, which gives the feather a larger open centre of ground colour.

Outer lacing Lacing around the outer edge of a feather as opposed to 'inner' lacing.

Parti-coloured Breed or variety having feathers of two or more colours, or shades of colour.
Paunch see 'Lobe'.
Pea comb A triple comb, resembling three small single combs joined together at the base and rear, but distinctly divided, the middle of one being the highest; best seen on the head of a well-bred Brahma.
Pearl eyed Eyes pearl coloured. Sometimes referred to as 'daw eyed'.
Pencilled spikes The spikes of a single comb that are very long and narrow; little broader at the base than at the top; generally a defect.
Pencilling Small markings or stripes on a feather, straight across in Hamburgh females (and often known as bands); or concentric in form, following the outline of the feather, as in the Brahma (dark), Cochin (partridge), Dorking (silver grey), and Wyandotte (partridge and silver pencilled) females, and fine stippled markings on females of Old English Game and brown Leghorns.
Peppering The effect of sprinkling a darker colour over one of a lighter shade.
Primaries Flight feathers of the wing, tucked out of sight when the bird is at rest. Ten in number.
Primary coverts see 'Flight coverts'.
Pullet A female fowl of the current season's breeding.
Pullet-breeder A term applied to birds, either male or female, selected to produce good standard bred pullets.
Pupil Black centre of eye.

Quill Hollow stem of feathers attaching them to the body.

Raspberry comb A comb somewhat resembling a raspberry cut through its axis (lengthwise) and covered with small protuberances.
Reachy Tall and upright carriage and 'lift' as in Modern Game.
Ring see 'Collar'.
Roach back Humped back, a deformity.
Rose comb A broad comb, nearly flat on top, covered with small regular points or 'work', and finishing with a spike or leader. It varies in length, width, and carriage according to breed.
Rust A patch of red-brown colour on the wings of females of some breeds, chiefly those of the black-red colour; brown or red marking in black fluff or breast feathers; known also as 'foxiness' in females.

Saddle The posterior part of the back, reaching to the tail of the male, and corresponding to the cushion in a female.
Saddle hackle see 'Hackles'.
Sandiness Giving the appearance of having been sprinkled with sand.
Sappiness A yellow tinge in plumage.
Scapulars Shoulder feathers, which may be elongated in waterfowl.
Secondaries The quill feathers of the wings which are visible when the wings are closed.
Self colour A uniform colour, unmixed with any other.
Serrations 'Sawtooth' sections of a single comb.
Shaft The stem or quill part of the feather.
Shafty Lighter coloured on the stem than on the webbing; a desirable marking in dark Dorking females and Welsummers. Generally a defect in other breeds.
Shank see 'Leg'.
Shank feathering see 'Feather legged'.
Sheen Bright surface gloss on black plumage. In other colours usually described as lustre.
Shell comb see 'Leaf comb'.
Shoulder The upper part of the wing nearest the neck feather. Prominent in Game breeds where it is often called the shoulder butt (see also 'Wing butt' and 'Scapulars').
Sickles The long curved feathers of a male's tail, usually applied to the top pair only (the others often being called the 'lesser' sickles), but sometimes used for the tail coverts.
Side sprig An extra spike growing out of the side of a single comb.
Single comb A comb which, when viewed from the front, is narrow, and having spikes in line behind each other. It consists of a blade surmounted by spikes, the lower (solid) portion being the blade, and the spaces between the spikes the serrations. It differs in size, shape and number of serrations according to breeds.
Slipped wing A wing in which the primary flight feathers hang below the secondaries when the wing is closed. This condition is often allied with split wing, in which primaries and secondaries show a very distinct segregation in many breeds of bantams.

Smoky undercolour Defective grey pigment in the undercolour of a bird.
Smut Dark or smutty colour where undesirable, such as in undercolour.
Soft feather Applied to breeds other than the hard feather group of Indian and Jubilee Game, Old English Game, Asil, Malay, and Modern Game.
Sootiness Grey or smokiness creeping in where it is not wanted, usually in undercolour.
Spangling The marking produced by a spot of colour at end of each feather differing from that of the ground colour. When applied to a laced breed, as the Poland, it means broader lacing at the tip of each feather. The spangle of circular form is the more correct, since, when of crescent or horseshoe shape, it favours the laced character.
Spike The rear leader on a rose comb.
Splashed feather A contrasting colour irregularly splashed on a feather.
Split comb The rear blade of a single comb is split or divided.
Split crest Divided crest that falls over on both sides.
Split tail Decided gap in middle of tail at base.
Split wing see 'Slipped wing'.
Sprig see 'Side sprig'.
Spur A projection of horny substance on the shanks of males, and sometimes on females.
Squirrel tail A tail, any part of which projects in front of a perpendicular line over the back; a tail that bends sharply over the back and touches, or almost touches, the head, like that of a squirrel.
Strain A family of birds from any breed or variety carefully bred over a number of years.
Strawberry comb A comb somewhat resembling half a strawberry, with the round part of the fruit uppermost; known also as the 'Walnut comb'.
Striping The very important markings down the middle of hackle feathers, particularly in males of the partridge variety.
Stub Short, partly grown feather.
Sub-variety see 'Variety'.
Surface colour That portion of the feathers exposed to view.
Sword feathered Having sickles only slightly curved, or scimitar shaped, as in Japanese bantams.
Symmetry Perfection of outline, proportion; harmony of all parts.
Tail coverts see 'Coverts'.
Tail feathers Straight and stiff feathers of the tail only. The top pair are sometimes slightly curved, but they are generally straight or nearly so. In the male fowl, main tail feathers are contained inside the sickles and coverts.
Tassel see 'Crest' and 'Beard'.
Tertiaries Feathers attached to the humerus (wing bone closest to the body) and overlying the secondaries. May be elongated in waterfowl.
Thigh That part of the leg above the shank, and covered with feathers.
Thumb-marked comb A single comb possessing indentations in the blade: a defect.
Ticked Plumage tipped with a different colour, usually applied to V-shaped markings as in the Ancona. Also small coloured specks on any part of feathers of different colour from that of the ground colour.
Tipping End of feathers tipped with a different coloured marking.
Top colour see 'Surface colour'.
Top knot see 'Crest'.
Tri-coloured Of three colours. The term refers chiefly to buff and red fowls, and is generally applied only to males when their hackles and tails are dark compared with the general plumage, and the wing bows are darker; a fault.
Trio A male and two females.
Triple comb see 'Pea comb'.
Twisted comb A faulty shaped pea or single comb.
Twisted feather The shaft and web of the feather are twisted out of shape.
Type Mould or shape (see 'Symmetry').

Undercolour Colour seen when a bird is handled – that is, when the feathers are lifted; colour of fluff of feathers.
Uropygium Parson's nose.

V-shaped comb see 'Horn comb'.
Variety A definite branch of a breed known by its distinctive colour or marking – for example, the black is a variety of the Leghorn. Sub-variety, a sub-division of an established variety, differing in shape of comb from the original – for example, the rose-combed black is a sub-variety of the black Leghorn. Thus the breed includes all the varieties and sub-varieties which would conform to the same standard type.
Vulture hocks Stiff projecting quill

feathers at the hock joint, growing on the thighs and extending backwards.

Walnut comb see 'Strawberry comb'.

Wattles The fleshy appendages at each side of base of beak, more strongly developed in male birds.

Web A flat and thin structure. Web of feather: the flat or plume portion. Web of feet: the flat skin between the toes. Web of wing: the triangular skin seen when the wing is extended.

Whiskers Feathers growing from the sides of the face (see 'Beard' and 'Muff').

Wing bar Any line of dark colour across the middle of the wing, caused by the colour or marking of the feathers known as the lower wing coverts.

Wing bay The triangular part of the folded wing between the wing bar and the point (see 'Diamond').

Wing bow The upper or shoulder part of the wing.

Wing butt or Wing point The end of the primaries; the corners or ends of the wing. The upper ends are more properly called the shoulder butts and are thus termed by Game fanciers. The lower, similarly, are often called the lower butts.

Wing coverts The feathers covering the roots of the secondary quills.

Work The small spikes or working on top of a rose comb.

Wry back A distorted bone structure usually causing a humpbacked condition.

Wry tail A tail carried awry, to the right or left side of the continuation of the backbone, and not straight with the body of the fowl.